MW00811556

NANOWEAPONS

NANOWEAPONS

A GROWING THREAT TO HUMANITY

LOUIS A. DEL MONTE

Potomac Books

AN IMPRINT OF THE UNIVERSITY OF NEBRASKA PRESS

Manufactured in the United States of America.

Library of Congress Cataloging-in-Publication Data

Names: Del Monte, Louis A., author. Title: Nanoweap-
ons: a growing threat to humanity / Louis A. Del
Monte. Description: Lincoln, NB: Potomac Books, an
imprint of the University of Nebraska Press, [2017] |
Includes bibliographical references and index.
Identifiers: LCCN 2016043320 (print) | LCCN
2016043925 (ebook)
ISBN 9781612348964 (cloth: alk. paper)
ISBN 9781612349121 (epub)
ISBN 9781612349138 (mobi)
ISBN 9781612349145 (pdf) Subjects: LCSH: Mil-
itary art and science—Technological innovations.
| Military weapons—Technological innovations. |
Nanotechnology—Risk assessment. Classification:
LCC U39 .D447 2017 (print) | LCC U39 (ebook) | DDC
355.8/2—dc23 LC record available at https://lccn.loc.
gov/2016043320
Set in Scala OT by John Klopping.

To my loving wife and lifelong partner, Diane Cuidera Del Monte

CONTENTS

Acknowledgments . . ix

Introduction . . xi

Part 1. The First Generation of Nanoweapons

1. What You Don't Know Can Kill You . . 3
2. Playing LEGOS with Atoms . . 17
3. I Come in Peace . . 29
4. The Wolf in Sheep's Clothing . . 45
5. The Rise of the Nanobots . . 79
6. The "Swarm" . . 91

Part 2. The Game Changers

7. The "Smart" Nanoweapons . . 101
8. The Genie Is Loose . . 111
9. Fighting Fire with Fire . . 123

Part 3. The Tipping Point

10. The Nanoweapons Superpowers . . 139
11. The Nano Wars . . 153
12. Humanity on the Brink . . 169

Epilogue . . 181

Appendixes . . 189

    1. Institute for Soldier Nanotechnologies

    2. Nanoweapons Offensive Capability of Nations

    3. The Events Leading to the Chernobyl Disaster

Notes . . 211

Glossary . . 225

Index . . 235

ACKNOWLEDGMENTS

I want to recognize the contributions of Diane Cuidera Del Monte. As a Renaissance woman, teacher, editor, and writer in her own right, Diane is capable of performing in numerous Arts mediums. As an intelligent wordsmith and art historian, she suggested editorial and structural changes to the book. She freely shared her talents, education, and insights on psychology, history, and human nature, which helped guide the completion of this work. She has always helped me dream larger and achieve greater. She thinks outside the box, outside the room that holds the box, and outside the house, which includes the room. She always refuses to be constrained by "artificial boundaries" and continually encourages our family and extended family to follow her lead.

I also want to thank my closest friend, Nick McGuinness, whose knowledge of history and the "real world" we live in astounds me. Finding a good friend is difficult. Finding a friend with knowledge and talent honed through "real-life" experience is extremely difficult. I believe it is rare to have a friend, like Nick, who will donate his time and talents to provide chapter-by-chapter editorial guidance, and I will forever be in his debt.

This book would never have found a publisher without the hard work and dedication of my agent, Jill Marsal, who is a founding partner of the Marsal Lyon Literary Agency. With her skillful guidance, we were able to construct a proposal that attracted the interest of several publishers. I found working with her a pleasure and an educational experience.

Lastly, I want to thank the excellent team at Potomac Books that worked to make this book available to the public, including Elaine Durham Otto, the freelance copyeditor.

INTRODUCTION

Nanoweapons are the most likely military weapons to render humanity extinct in this century. This is not a philosophical issue. This is about whether you and yours will survive through this century. Having made such a dire assertion, you may wonder if I am being an alarmist. Consider this. The events that most people consider likely to cause humanity's extinction, such as a large asteroid impact or a super-volcanic eruption, actually have a relatively low probability of occurring, in the order of 1 in 50,000 or less. In 2008 experts surveyed at the Global Catastrophic Risk Conference at the University of Oxford suggested a 19 percent chance of human extinction by the end of this century, citing the top four most probable causes:

1. Molecular nanotechnology weapons: 5 percent probability

2. Superintelligent AI: 5 percent probability

3. Wars: 4 percent probability

4. Engineered pandemic: 2 percent probability

Obviously nanoweapons are at the top of the list, having a 1 in 20 probability of causing human extinction by the end of this century. Notice that biological weapons (item 4), which have been a mainstay apocalyptic theme in both fiction and nonfiction, come in as a distant fourth, with only a 1 in 50 probability of causing human extinction.

Almost every book on nanotechnology speaks to the enormous benefits it will yield. Few mention the inherent risks associated with nanotechnology. Even fewer mention nanoweapons. You may ask, Why do books on nanotechnology rarely ever mention nanoweapons? The answer is that nanoweapons are "classified," meaning the technology and its military applications are either "Secret" or "Top Secret." The technologists involved in their development cannot publish their research in scientific journals, speak about it at scientific conferences, or give media interviews about it.

It is one thing to read a scholarly Oxford study that proclaims nanoweapons threaten human extinction. Philosophers often ponder esoteric issues that never come to fruition. Thus it is reasonable to ask, Just how real are nanoweapons? Let us address this question using a time-tested technique, namely, "follow the money." In 2000, under President Bill Clinton, the U.S. government launched the National Nanotechnology Initiative (NNI), a research and development initiative involving the nanotechnology-related activities of twenty-five federal agencies with a range of research and regulatory roles and responsibilities. Since its inception, the government has allocated over $20 billion to developing nanotechnology. The actual amount that is spent on nanoweapons remains Top Secret. However, based on publicly published budget allocations, it is reasonable to estimate that anywhere between a third to half of all nanotechnology-based materials and systems under NNI are destined for military application. Indeed, several are already deployed and deemed combat ready. The initial examples I provide in chapter 1 will sound like they are right out of *Star Trek*. Factually, a new arms race is under way. Nanowerk.com, the leading nanotechnology portal, reports, "All major powers are making efforts to research and develop nanotechnology-based materials and systems for military use." Based on publicly available information, China, Russia, and the United States are competing in a multibillion dollar nanoweapons arms race. Other nations, like Germany, are close on their heels. A new paradigm fuels this race. The superpowers of the future will be those nations with the most capable nanoweapons. Given the above facts, the existence of nanoweapon development and deployment is beyond dispute, but it is appropriate to ask, Why do nanoweapons threaten the survival of humanity? The simple answer is "control." Controlling nanoweapons is as problematic as controlling biological weapons.

Let us understand the control issue using an example. Assume one nation develops artificially intelligent nanobots, tiny robots about the size of insects, capable of numerous military missions from surveillance to assassination. The size of nanobots makes them easy to transport and difficult to detect. In addition, by mid-century, current nanotechnology projections suggest artificially intelligent self-replicating nanobots will become a reality. These nanobots are capable of replicating themselves by literally seeking the right atoms and assembling a clone. In effect, they are the technological equivalent of biological weapons. Imagine such technology in the hands of a rogue state or terrorist group. Weaponized self-replicating smart nanobots would represent the ultimate doomsday device. Once released, their mission would be twofold: kill humans and replicate. Assuming the self-replicating smart nanobots are equivalent to a deadly biological pandemic, 90 percent of the human race could fall victim to their attack in a matter of weeks. All this may sound like science fiction, but it is not. This will become evident in the chapters that follow.

Nanoweapons will describe this new class of military weapons in layperson prose. Assertions are annotated, making their source apparent. To assist comprehension, please consult the glossary. This book will enable you to trace the emergence of nanoweapons from concept to current deployment. It will discuss the nanoweapons in development and close to deployment. It will project the nanoweapons most likely to dominate the future battlefield in the second half of this century. Most important, it will raise the control issue by addressing the question, Will it be possible to develop, deploy, and use nanoweapons in warfare without rendering humanity extinct?

Many books offer concerns about technology, but provide no actionable plan. This book will propose strategies that could form the basis to ensure nanoweapons do not render humanity extinct.

I do not claim to have all the answers. But the pages that follow will clearly define the threat. I also believe the proposed strategies to address the threat are worthy of consideration.

I admit my forecasts on nanoweapon timing may be off, perhaps by as much as a decade. The development of nanoweapons continues under a cloak of secrecy by every nation involved. This makes delineating an exact timeline challenging. However, by researching the available information, it is possible to connect the dots to make an educated guess regarding when specific classes of nanoweapons will become available. Let me provide an example. When I was a boy, I enjoyed building small plastic models, especially military models. In 1954 the U.S. Navy launched the USS *Nautilus* (SSN-571), the world's first atomic submarine. Because it made headlines, it became a common topic of conversation. As a model builder, I wanted to build it, as did many other hobbyists. Revell Hobby Kits was aware of the demand for a plastic model build kit and quickly offered one. The Revell model of the USS *Nautilus* was apparently accurate, pinpointing the almost exact location of the atomic reactors on the *Nautilus*, which at the time was classified Top Secret. This started some people speculating that Revell had the Navy's secret plans. Apparently the Revell's model engineers knew how to piece various elements of their research together even without the Internet. This childhood memory impressed me so deeply that it is still with me to this day. My friends and I thought it a joke. Obviously the U.S. Navy likely found it frustrating at the very least. My point is simple. Collating available information can yield surprisingly accurate insight into classified information. Using this approach, let us peek under the cloak of secrecy to see the nanoweapons that currently exist and those likely to exist in the coming decades. Let us understand how they threaten humanity's survival. Most important, let us consider what strategies are necessary to ensure they do not become our undoing.

NANOWEAPONS

Part 1

THE FIRST GENERATION OF NANOWEAPONS

1   What You Don't Know Can Kill You

> Reports that say that something hasn't happened are always interesting to me, because as we know, there are known knowns; there are things we know we know. We also know there are known unknowns; that is to say we know there are some things we do not know. But there are also unknown unknowns—the ones we don't know we don't know. And if one looks throughout the history of our country and other free countries, it is the latter category that tend to be the difficult ones.
>
> —DONALD RUMSFELD

Scenario of a Nanoweapons Attack—An enemy could kill you before you finish this sentence. The enemy need not be present nor the killer human. You are a civilian, not a military combatant. You are not associated with any political action group. In fact, you cracked open this book for a quick read while finishing your morning coffee and getting ready for work. You're employed as a chef for a nursing home, nothing strategic to the national defense. Your political beliefs are a conglomeration of the previous week's news cycles. Friends say you are easygoing and likable. You have a loving fiancée whom you are marrying next month. The calendar on your refrigerator is your "to do" list, surrounded by business cards from the wedding photographer, florist, caterer, and such. You have plans for a honeymoon and a family. Yet for some unknown reason you are one of the first victims of a nanoweapons attack. There will be no way to see it coming and no way to escape. Your autopsy will show that you died from an "unknown cause." Some at your funeral will mourn and openly ask, Why?

Their grief will be short-lived and their question quickly answered. Many of them will meet the same fate within days. Your death was only the beginning. In fact, numerous people like you are dying in major cities around the world. If you live in the United States, the Centers for Disease Control and Prevention may just be beginning to understand the nature of the attack. Some high-ranking government officials know the United States is under attack. They and their families are at secure locations to ensure continuity of government. The president and his family have been in the White House bunker for several days. The enemy responsible for the attack remains unknown. All national resources are working to find a way to neutralize the nanoweapons attack. The president is getting ready to address the nation and launch a counterattack. Every branch of the U.S. military is at DEFCON 2, signaling that a nuclear counterattack is imminent. U.S. submarines are strategically scattered throughout the world's oceans. A third of all U.S. strategic bombers are airborne 24/7. Every missile silo stands ready to launch. NORAD is sealed.

While Americans await the president's address, the chill of dread is in the air. Schools close, and the streets are empty. Television news coverage has devoted more airtime to the "mysterious deaths" than to any news story in TV history, with a bombardment of cold statistics and "expert" commentary from every major network for the past four days. The Emergency Broadcast System interrupts repeatedly to advise all citizens to stay indoors and avoid contact with others to whatever extent possible. The shelves of food markets and gun shops are bare, victims of hoarding and looting. Three days ago, the president suspended all unnecessary government services, like the U.S. Post Office and the Internal Revenue Service. The National Guard is maintaining order and keeping hospitals operational as daily tides of new victims arrive who often die before a bed becomes available. Finally, from the White House bunker, the president addresses the nation.

"My fellow Americans," he begins with a solemn demeanor, "I regret to report that nine days ago the United States and many of its NATO allies, as well as Russia and China, suffered attacks from unknown forces using a new type of weapon. This new weapon is similar to a biological weapon, but it is not biological. It is technological. We describe it as a 'nanoweapon.' Our best scientists are working to put an end to the mysterious deaths. I am confident we will do so, and those responsible for the attacks will face justice. As I speak, the combined forces of the United States and its NATO allies, along with Russia and China, are launching a full counteroffensive against all nations and terrorist groups we deem enemies of humanity. We act with resolve that our nations will not perish from the Earth."

The president pauses, taking a deep breath, looks directly into the camera, and continues: "In this time of emergency, I urge all to remain calm, with goodwill toward each other. Let those who can help reach out to those less fortunate. I pledge that our offensive will mete justice and cripple any adversary's ability to continue their attacks. We are past the point of diplomacy. Those we deem enemies of humanity and responsible for, or in support of, these attacks forfeit their right to share planet Earth. Tomorrow they will be gone. Tomorrow we will prevail. God bless the United States of America."

Concurrent with the president's two-minute national television address, the most devastating counterattack in history begins. Unfortunately, the counterattack has to target all possible perpetrators, in other words, a broad-spectrum attack against every suspected adversary in an effort to thwart the release of more nanoweapons on the United States and its allies. For the first time in nearly a century, nuclear weapons have unleashed, as the book of Revelation put it, "the lake of fire" on Earth. Countless millions die, the innocent along with the guilty. Tomorrow and the future of humanity have become questionable.

Although the preceding scenario is fictitious, it is entirely plausible. Nanoweapons are real and a new arms race is under way. Based on publicly available information, China, Russia, and the United States are competing in a multibillion dollar nanoweapons arms race. Other nations, like Germany, are close on their heels. A new paradigm fuels this race. The superpowers of the future will be those nations with the most capable nanoweapons. This is easy to illustrate. Recall the first sentence of this chapter: "An enemy could kill you before you finish this sentence." Here is how such a nanoweapons attack could happen. Assume one nation develops artificially intelligent nanobots, with functionality similar to mosquitos. Also, assume the nanobots are capable of seeking and injecting toxin into another nation's humans. The smallest known flying insects are fairyflies, belonging to the family of chalcid wasps. Fairyflies are approximately 139 microns long (139 millionths of a meter). This suggests a plausible size for a lethal nanobot. If the toxin is botulism, the human lethal dose is 100 nanograms. If we assume the toxin payload each nanobot carries is 1,000 nanograms, similar to the weight ratio of a fighter aircraft to its ordinance payload, each nanobot could theoretically kill ten humans. An autopsy will reveal the presence of botulism and may attribute the death to food poisoning, not foul play. Even worse, if it is botulinum toxin type H, the most deadly in existence, there is no known antidote. Once injected, it becomes only a matter of days before your brain shuts down and you die. Most medical examiner labs are unfamiliar with botulinum toxin type H and not able to detect it. The injection point would be invisible to conventional autopsy techniques. This means that it is entirely possible that the medical examiner will attribute your death to an unknown cause, but not suggest foul play. The actual injection could take place within seconds. You may not be aware of it. You may never have heard of nanoweapons and botulinum toxin type H. It does not matter. Once injected, you are going to die.

Above, we discussed mosquito-like nanobots. They do not exist now, but the technology to build such a nanoweapon is only one or two decades away. No nation has a defense against such a nanoweapon. You may think this is far-fetched, but the idea of poisoning someone with a nearly imperceptible device is not new. A well-documented case involves Georgi Markov, novelist, playwright, and broadcast journalist for the BBC World Service. As a Bulgarian dissident, Markov was critical of the incumbent Bulgarian communist regime under Chairman Todor Zhivkov. Because of his criticism, many speculate that the Bulgarian government decided to silence him. On September 7, 1978, Markov walked across Waterloo Bridge spanning the river Thames. While waiting to take a bus to his job at the BBC, he felt a sharp pain in the back of his right thigh. He described the pain as a bug bite or sting. The pain caused Markov to look behind him. He saw a man picking up an umbrella and hurriedly crossing the street, where he got into a taxi and sped away. After arriving at the BBC World Service offices, Markov noticed a small red pimple had formed at the site of the sting, which continued to cause pain. He told one of his colleagues at the BBC about this incident. That evening he developed a fever and sought treatment at St. James' Hospital in Balham. He died on September 11 at the age of forty-nine. Due to the suspicious circumstances, the Metropolitan Police ordered an autopsy, which revealed a spherical metal pellet the size of a pinhead embedded in Markov's leg. The pellet had two holes drilled through it, producing an X-shaped cavity, which showed traces of ricin. A sugary substance coated the tiny holes, trapping the ricin inside. Once the pellet was injected into Markov's body, the sugary coating melted and the ricin found its way into his bloodstream. At the time, there was no known antidote to ricin. The intelligence communities term this event the "Umbrella Murder." In this case, a tiny pellet carried the toxin. Although this is arguably much larger than mosquito-like nanobots, it demonstrates a significant point. Something extremely small, with an almost miniscule amount of toxin, can kill a human.

If you imagine 50 billion mosquito-like nanobots, each carrying 1,000 nanograms of botulinum toxin H, released into the world's population, it is easy to understand that nanoweapons could represent a threat capable of rendering humanity extinct. Even more frightening, these nanobots could be carried in a suitcase.

Current nanotechnology projections suggest that by 2050 artificially intelligent self-replicating nanobots will become a reality, designed and manufactured by superintelligent computers. We will discuss this further in a later chapter. However, at this point, it is important to understand that nanoweapons are not science fiction. In fact, the United States is already deploying nanoweapons, and their use in warfare is only a matter of time. However, we may be getting a bit ahead of ourselves. Let us start at the beginning.

If you have never heard of nanoweapons, you are in the majority. Most people have never heard the words "nanotechnology" or "nanoweapons." Even in technically advanced countries, like the United States, the majority of adults are unaware that nanoweapons even exist. Is this an exaggeration? No! Just look at the facts.

Fact 1: The National Nanotechnology Initiative intentionally omits any mention of nanoweapons in its mission and goals, but it allocates a significant portion of its budget to their development.

In 2000, under President Bill Clinton, the U.S. government launched the National Nanotechnology Initiative (NNI), a research and development initiative involving twenty-five federal agencies with a range of research and regulatory roles and responsibilities. Its mission and goals suggest no role in developing nanoweapons, but its budget allocations tell a different story. According to public records, in 2015 Department of Defense (DoD) programs accounted for more than 11 percent of the NNI budget. However, this excluded funding to agencies like the Defense Advanced Research Projects Agency, which also develops nanoweapons. In my judgment, the actual funding into nanoweapons is likely Top Secret or, at a minimum, Secret.

Let us be clear on what it means when information is classified. Generally, Top Secret applies to information that, in the wrong hands, could cause grave damage to national security. An example would be our nuclear launch codes. Secret applies to information that could cause serious damage to national security. An example would be fabrication techniques used to make radiation-hardened integrated circuits. These are circuits able to withstand high radiation, typically associated with a nuclear explosion. DoD applications include communication satellites and strategic missiles, since both are likely to experience high radiation exposure during a nuclear war.

I held a Secret clearance during my work on DoD programs at Honeywell. This meant that I could have access to information classified as Secret, but only if I had a "need to know." Having a Secret clearance did not allow me open access to all information classified Secret. I had to have a need to know the information in order to perform my work. This may mean that even the president of the United States and members of Congress may not know how much money is going into nanoweapons. It is also likely that technologists working on nanoweapons have at least a Secret clearance. This means they cannot publish their research on nanoweapons in a scientific journal, speak about it at conferences, or give media interviews. Given all the secrecy that surrounds nanoweapons, it is not surprising that most people have never heard of them.

If you recall, early combat use of modern stealth aircraft only came to the public's attention in December 1989 during Operation Just Cause in Panama and later in 1990 during Operation Desert Shield to liberate Kuwait. However, the actual engineering to create stealth aircraft started in 1975, when engineers at Lockheed Skunk Works found that an aircraft made with faceted surfaces could have a very low radar signature because the surfaces would radiate almost all of the radar energy away from the receiver. Their inception to deployment spanned about fifteen years. If the need to deploy stealth aircraft had not emerged, their existence would still be classified Top Secret and the public would be in the dark.

Fact 2: There is little to no public information regarding nanoweapons. For example, the 2007 *New York Times Almanac*, claiming to be "the world's most comprehensive and authoritative almanac," did not include the word "nano" in its thirty-page index. A Google search on March 24, 2016, using the keyword "nanoweapons" yielded 10,800 search returns. That may seem like a lot of information, but in the larger scheme, it suggests there is scant information available. A second Google search using the keyword "nuclear weapons" yielded over 14,000,000 search returns. The amount of information on nanoweapons is .07 percent that of nuclear weapons. Many may attribute this difference to the relatively recent emergence of nanoweapons. To some extent, that is true. However, the U.S. government has been pursuing nanoweapons since 2000. Clearly it is not only the age of the technology causing the disparity of information. It is the secrecy. Unlike nuclear weapons or even stealth aircraft, the United States has not deployed nanoweapons in combat. As a result, they remain secret and garner little press coverage.

Given these facts, it is not surprising that a national poll of U.S. citizens in 2007 revealed 79 percent had not heard about nanotechnology. A Harris Poll of 2,467 U.S. adults in 2012 found over 60 percent had never heard of nanotechnology. The simple fact is that even today it is likely that most Americans are not aware of nanotechnology, let alone nanoweapons.

At this point you may wonder, What started the nanoweapons arms race? Like most things in science, nanoweapons started with a concept. Physicist and Nobel laureate Richard Feynman's talk "There's Plenty of Room at the Bottom" introduced the concept at the American Physical Society meeting at the California Institute of Technology on December 29, 1959. Although Feynman never used the words "nanotechnology" or "nanoweapons," he described a process in which scientists would be able to manipulate individual atoms and molecules. At the time, no such process existed.

Inspired by Feynman's talk, engineer Kim Eric Drexler popularized the term "nanotechnology" in his 1986 landmark book, *Engines of Creation: The Coming Era of Nanotechnology*. In 1991, while Drexler was at MIT, his doctoral thesis became the foundation for another book, *Nanosystems: Molecular Machinery, Manufacturing, and Computation*.

It is fair to argue that Feynman and Drexler are the fathers of nanotechnology. In many ways, though, their prophetic visions were ahead of the prevailing scientific dogma. In 2001 Nobel laureate Richard Smalley criticized Drexler's work as naïve in a *Scientific American* article, "Of Chemistry, Love, and Nanobots," and subtitled "How soon will we see the nanometer-scale robots envisaged by K. Eric Drexler and other molecular nanotechnologists? The simple answer is never." Smalley made several technical arguments that Drexler refuted point by point as a "straw-man attack." Unfortunately, this is normal in the scientific community, when the proposed science stretches beyond the limits of the known science. Details aside, history favors Drexler. Point of fact, nanotechnology exists. Unfortunately, nanoweapons also exist, and even more are under development.

What do we really know about nanoweapons? Surprisingly, from the mountain of information on nanotechnology, one in every 2,500 pages describes the use of nanotechnology for military applications. However, be advised. The information in the public domain is likely one to two decades old. In the interest of clarity, let us examine several examples.

In 2007 the Russian military successfully tested the world's most powerful nonnuclear air-delivered bomb, nicknamed the "father of all bombs." Even though it only carries about seven tons of explosives compared with more than eight tons of explosives carried by the U.S. Massive Ordnance Air Blast bomb, nicknamed the "mother of all bombs," the Russian bomb is four times more powerful because it uses nanotechnology-enhanced explosives. In a counterpunch, the U.S. Department of Defense demonstrated the feasibility of creating compact bombs that use nanometals, such as nanoaluminum, to create explosives more powerful than conventional bombs. Since the deployment of such bombs is classified, we do not know if the United States has a bomb equal to or even greater than Russia's. However, the probability of the United States having such a bomb is a near certainty.

In 2013, the U.S. Navy announced the deployment of a laser weapon on the USS *Ponce*. While laser weapons are not a new con-

1. Nanotechnology-based laser weapon. *America's Navy*. U.S. Navy photo by John F. Williams.

cept, since their pursuit traces back to the 1950s Cold War, this one ignited the public's imagination, especially Trekkies, familiar with futuristic weapons like phasers. Similar to *Star Trek's* phasers, the new laser weapon is being refined to provide a wider range of tactical options, such as limiting the damage to a targeted aircraft and stunning an enemy combatant versus destroying the enemy aircraft and killing an enemy combatant. Although the laser weapon's technology is secret, numerous articles argue that the rapid development of nanotechnologies over the last decade enabled significant internal component improvements of the solid-state laser system, making it deployable as a weapon. Figure 1, released by the U.S. Navy, shows the laser weapon on the uss *Ponce*.

To my mind, the most horrific near-term nanoweapons are mini-nukes. These nanoweapons are in the development phase by Russia, Germany, and the United States. While the exact technology is secret, much of the science that underpins mini-nukes is in the public domain. For example, a high-power laser could

WHAT YOU DON'T KNOW CAN KILL YOU

trigger a small thermonuclear fusion explosion, using a mixture of tritium and deuterium. Nanotechnology could make the laser and fusion materials extremely small, able to fit easily into a jacket pocket, and weigh only five pounds. The blast from such a bomb would range from one ton to a hundred tons of conventional explosives. The nuclear fallout would be negligible. Technically these are not weapons of mass destruction but an entirely new category. The amount of fissionable material is minuscule, so the detection of a mini-nuke would be extremely difficult. Since their blast is limited and the radioactive fallout negligible, their use on the modern battlefield would be compelling. In theory, their delivery could be accomplished via the delivery systems we currently deploy, such as cruise missiles. The mini-nukes are likely to make using nuclear-type weapons in warfare a reality.

The above represents just a few recent examples of nanoweapons. We will cover more throughout this book. However, ask yourself this question, Have you ever heard of Russia's "father of all bombs," nanoweapon lasers, and mini-nukes? I suspect, based on previous polls, that most people have not, but make no mistake about their lethality. The almost 250,000 Japanese who died when atomic bombs dropped on Hiroshima and Nagasaki never heard the phrase "atomic bomb." Their best scientists did not know that the energy of atoms could be unleashed to cause such devastating destruction. Factually, history teaches us that ignorance is not a shield against horrific weapons.

How horrific are nanoweapons? Consider the simplest of all, the nanoparticle. This is an object, between 1 and 100 nanometers in diameter, that behaves as a single entity with respect to its properties. Nanoparticles are widely used in detergents, cosmetics, electronics, optical devices, medicine, and food packaging materials. How can they become a weapon? The simple answer is some nanoparticles are highly toxic. We do not fully understand the full effect of nanoparticle exposure on people and animals, but we do know that the size of nanoparticles allows living tissue to absorb them more readily than other known toxins.

Nanoparticles are able to cross biological membranes and access cells, tissues, and organs that their larger counterparts cannot. Alarmingly, there is no authority to regulate nanotechnology-based products, including nanoparticles, but there is significant scientific evidence that some may be usable as weapons. Clearly a nation intent on developing nanoweapons could turn its attention to developing highly toxic nanoparticles. Depending on the properties of the toxic nanoparticles, they may cause irreparable harm when inhaled or eaten. This means an enemy could potentially kill millions of people and animals by releasing toxic nanoparticles into a nation's reservoirs, its environment, or somewhere along a nation's bio food chain. Unfortunately, some toxicity studies indicate that considerable exposure is necessary before the onset of symptoms. Thus it might take weeks or months for the first toxic nanoparticle symptoms to surface. To make matters worse, the current detection of nanoparticles requires expensive and complex analytical instruments. Even when the first symptoms surface, it might take days, weeks, or even months to determine that the culprit is a toxic nanoparticle. By the time initial symptoms appear and the Centers for Disease Control and Prevention or the World Health Organization makes a definitive diagnosis, the bulk of a nation's population could already have absorbed a lethal dose and be beyond treatment.

In later chapters, we will discuss complex nanoweapons. However, I thought it instructive to discuss toxic nanoparticles here for four reasons:

1. Exposure to nanoparticles is already occurring, resulting from their widespread commercial use. Similar to asbestos and mercury poisoning, the toxicity of some nanoparticles over time could produce irreparable harm to humans and animals.

2. It demonstrates the ease of producing a nanoweapon. This nanoweapon does not require a complex molecular system. It

simply requires toxic nanoparticles, which some nations already know how to produce.

3. It demonstrates the ease of delivering the nanoweapon to the intended target. One suitcase filled with toxic nanoparticles could be sufficient to wipe out a major city.

4. It makes the point that the horrific effects of toxic nanoparticles are largely unknown and specific to the toxic nanoparticle. However, the highly concerning part is that nanoparticles do not have to be lethal to cause irreparable harm. One study suggests nanoparticles may be able to suppress a person's appetite to the point where they stop eating altogether. In humans, this could manifest itself as malnutrition. The symptoms of malnutrition are numerous, but most worrying are the mental effects, such as depression, irritability, and dizziness. A nation's military with these symptoms could become dysfunctional.

Ironically, the next big things in advanced military weapons will be small, essentially invisible. Their development proceeds under the premise of making us more secure. However, the reality is just the opposite. Their emergence threatens human extinction. As we witness the emergence of nanoweapons, I consider what American physicist and the scientific director of the Manhattan Project, Robert Oppenheimer, said in an interview about the Trinity atom bomb test:

> We knew the world would not be the same. A few people laughed, a few people cried, most people were silent. I remembered the line from the Hindu scripture, the *Bhagavad-Gita*. Vishnu is trying to persuade the Prince that he should do his duty and, to impress him, takes on his multi-armed form and says, "Now I am become Death, the destroyer of worlds." I suppose we all thought that, one way or another.

Nanoweapons, similar to nuclear weapons, are game changers. However, they promise to be even harder to control and more lethal than nuclear weapons.

2 Playing LEGOS with Atoms

The principles of physics, as far as I can see, do not speak against the possibility of maneuvering things atom by atom.

—RICHARD FEYNMAN

Manipulating atoms and molecules like LEGO blocks may sound playful, but that is what nanotechnology engineering is about. How is this even possible? Well, to be honest, it was not possible until relatively recently. However, nature has been doing nanotechnology engineering for billions of years. In her slow evolutionary way, Mother Nature was the first "scientist" to use nano processes to manufacture nanoparticles, nanostructures, and nanosystems. Let us consider some simple examples.

If you stand on the beach of the Atlantic Ocean, you can smell the ocean. Natural processes, like sea spray, create salt nanoparticles that enter our nostrils, and we are able to smell them. If you walk along the beach and find an abalone shell, you will have discovered another of Mother Nature's engineering marvels. Even though the abalone shell is 98 percent calcium carbonate, its nanostructure makes it 3,000 times stronger than rocks with the same chemical composition. The shell was created layer by layer in sheets about 50 to 200 nanometers thick.

Consider the mussels that cling to rocks. When I was a youngster living in New Jersey, my grandparents would often take me to the beach, and there they would harvest mussels from the man-made rock jetties projecting into the ocean. Have you ever tried to pull a mussel from a rock? They are essentially "superglued" in place and difficult to pry loose. My grandparents did it with a screwdriver and a lot of muscle. Riding home on the train with a paper bag filled with mussels, my mind only focused on the spaghetti with mussels we would have for dinner. I never thought about how the mussels were able to glue themselves to submerged rocks. But now I know that the glue used by the mussel is another one of nature's nanotechnologies. In effect, when a mussel uses its foot to attach to a rock, it releases molecular bubbles, which we could term "nanobubbles," that have an even smaller amount of adhesive in the center of each bubble. A shell of hydrophilic (water-loving) atoms called a micelle surrounds each nanobubble. The nanobubble protects the adhesive until it reaches the rock. When it reaches the rock, the bubble pops, deposits the adhesive on the rock, and glues the mussel in place.

Nature's nanotechnology engineering goes well beyond the ocean. For those that live inland and have never had the opportunity to explore an ocean beach, the marvels of nanotechnology engineering are all around you. For example, if you live in a rural area and walk through a pine forest during the summer, you will experience a fresh evergreen scent. The trees' resins create terpene, which is a natural nanoscale hydrocarbon (an organic compound consisting entirely of hydrogen and carbon). Remarkably, our noses are sufficiently sensitive to smell these nanoscale molecules, and most people even like the scent.

As you approach a pond, you are likely to see frogs jumping from the foliage into the water. How did they know you were coming? This is another one of nature's most marvelous nanosystems. If you have ever been on a freestanding balcony, you were on a structure known as a cantilever. A cantilever is simply a rigid structural element anchored at only one end. The freestanding balcony is a protruding cantilever. An important mechanical property of a protruding cantilever is that when subjected to a structural load, such as a person on a balcony, all the stress transmits to its anchor point. This makes the anchor point highly sensitive to the structural load. This is what is going on inside the frog's ear. The frog's inner ear has nanomechanical cantilevers, each known as a "haircell sacculus," that move as little as 3 nanometers in reaction to the sound of our footsteps as we approach them.

Even if you live in a big city like Chicago, you will still experience nature's nanotechnology. For example, there are butterflies. The intricately colorful wings of a butterfly are the result of nanoparticles that act as a photonic crystal, an optical nanostructure fine enough to interfere with visible light and produce a color related to its surface property. If you take a walk through a city park, you will experience another of Mother Nature's nanotechnology engineering feats, namely flowers. For example, let us say the park contains nasturtium flowers.

The leaves of the nasturtium flower, like the lotus flower, have a nanostructured surface. It is like closely packed microscopic fuzz. It acts to keep water droplets from adhering to the surface. As a result, the water droplets run off, taking surface dirt with them. This process is termed the "lotus effect." The purpose in keeping the nasturtium leaves clean is to enable the plant to carry on the process of photosynthesis efficiently. Photosynthesis is nano process, meaning it occurs at the atomic and molecular level. Photosynthesis allows green plants and some microorganisms to use sunlight as a power source to extract carbon dioxide and water and then rearrange the atoms to form oxygen and carbohydrates. Without the nano process of photosynthesis and the oxygen it creates, no animal life would have evolved on Earth.

We humans also operate using nano processes. A strand of DNA that is found in every cell of our bodies is only 2 nanometers in diameter. Yet it carries the genetic instructions for the development, function, and reproduction of our bodies. In fact, all living organisms, with the exception of some viruses, rely on DNA for their genetic code. However, nature goes far beyond the use of nanoscale DNA molecules when it comes to engineering humans. For example, nanoscale proteins form a large part of the structure of the human body and numerous other animals, from muscle tissues to antibodies.

When Mother Nature develops a successful nano process, she tends to use it widely. For example, micelles are found inside our bodies. In order for our bodies to absorb vitamins like D and E, they must travel in micelles, similar to how mussel nanobubbles travel in water. Nano processes within our bodies do not stop with the delivery of vitamins. The most critical element our bodies need is oxygen. Without oxygen, we would die within three minutes. Every human knows, even in the most primitive societies, that breathing is critical to staying alive. This is how we get oxygen into our bodies, but it is not the whole story. Once inside our bodies, hemoglobin (a red protein less than 10 nanometers in size) carries the oxygen to every cell in our bodies via our bloodstream. Without this nano process, we and other vertebrates would cease to exist.

There is little doubt that Mother Nature is unrivaled when it comes to nanotechnology engineering. She had a three billion year head start on human scientists, and even today our best nanotechnology engineering is still unable to equal her numerous accomplishments. This is not for lack of trying. One approach to nanotechnology engineering is to attempt to duplicate the nano processes we see in nature. For example, we have tried to arrange the atoms of calcium carbonate to duplicate the strength found in abalone shells, but with little success. We have made remarkable progress in this area. However, we have only achieved one-tenth of its strength.

These examples are just a small sample of the nanoparticles, nanostructures, and nanosystems found in nature. I offer them to make a significant point. Nanotechnology engineering is real. It existed long before the evolution of humans. It is the foundation for all life on Earth.

At this point, it is obvious that nature has mastered the ability to play LEGOS with atoms. However, it is fair to ask, How do humans create nanoparticles, nanostructures, and nanosystems?

You can visualize our development of nanotechnology as the result of a "perfect storm" in science. Three critical events came together in a relatively short period, 1981–89. When the storm cleared, the research fields of nanotechnology emerged. These are the three events:

1. The scanning tunneling microscope (STM) was invented in 1981. This allowed individual atoms to be "seen" for the first time in history. Work in the early twentieth century, especially in the field of quantum mechanics, made a convincing case that atoms existed and provided insight into their behavior. However, the STM provided an unprecedented visualization of their existence. Its developers, Gerd Binnig and Heinrich Rohrer at IBM Zurich Research Laboratory, received a Nobel Prize in physics in 1986.

2. In 1986 K. Eric Drexler published his landmark book, *Engines of Creation: The Coming Era of Nanotechnology*. In it Drexler suggested controlling atoms to build nanoscale machines. Sparked by Drexler's conceptual framework and the invention of the STM, the prospect of atomic control gleaned serious widespread attention.

3. In 1989 Don Eigler, an IBM physicist, used the STM tip to manipulate individual atoms. This was a historic first. It demonstrated that it was possible to use the STM not only to visualize atoms but to move and place them. He spelled out the letters "IBM" using 35 Xenon atoms on a nickel surface.

Beginning in the early 2000s, the fields of nanotechnology began to emerge. Notice that I use the plural "fields." That is intentional. Recall the National Nanotechnology Initiative definition we discussed in chapter 1: "Nanotechnology is science, engineering, and technology conducted at the nanoscale, which is about 1 to 100 nanometers." This definition finds wide acceptance within the scientific community, but let me clarify one point. This definition does not require that macroscale products (the typical size of everyday products) only be produced by a gradual buildup, atom by atom or layer by layer, from the nanoscale, similar to the way nature builds the abalone shell. The NNI requires only one dimension of the macroscale product to be in nanoscale, 1 to 100 nanometers, for the product to fit within the category of nanotechnology. This interpretation opens the door for numerous scientific fields to engage in nanotechnology research and application, including the fields of surface science, organic chemistry, molecular biology, semiconductor physics, and microfabrication. It is common to see the plural "nanotechnologies," referring to the broad range of research and applications that share only one commonality: size.

If you look closely at the emergence of nanotechnology in the 2000s, numerous events demonstrate how seriously some members of the scientific community took this new research category. In addition to an increase in scientific interest, there was considerable political and commercial interest. Here are some of the key events:

> Scientifically: In the early 2000s, a number of heavy hitters in the scientific community, including Drexler, Nobel laureate Richard Smalley, chemist George M. Whitesides, and computer scientist Ray Kurzweil, began to debate the viability of nanotechnology to achieve the vision forwarded by Drexler. In his book *Engines of Creation*, Drexler argued that nanotechnology would eventually fuel the ability to build a nanoscale "assembler" that would be capable of building a copy of itself or any complex system via an atom-by-atom approach. Many refer to this approach as "molecular manufacturing." The debate was theoretical and mainly increased scientific interest in nanotechnology. In 2004 Britain's Royal Society and the Royal Academy of Engineering published "Nanoscience and Nanotechnologies: Opportunities and Uncertainties." The report listed both the opportunities and potential risks associated with nanotechnology. In particular, it singled out nanoparticles as a potential

health hazard and called for regulation. The theoretical nanotechnology debate and the Royal Society's report had little impact on slowing the progress of nanotechnology. Governments and industries quickly realized the potential of nanotechnology and acted accordingly.

Politically: Governments began to fund and direct research in nanotechnology. In 2000 the United States formed the National Nanotechnology Initiative and began spending billions of dollars on research. Europe pursued nanotechnology research through their European Framework Programmes for Research and Technological Development (Framework Programmes), a collaborative organization made up of European Union member nations. China began pursuing nanotechnology in 2000 and funded its development second only to the United States. Similar to the United States, a significant portion of their funding focuses on weapons. In 2007 Russia officially began nanotechnology research with the passage of a law that resulted in the creation of the Russian Corporation of Nanotechnologies. Russia's efforts in nanotechnology have been rocky at best, due to financial constraints and corruption.

Commercially: A large number of nanotechnology-based commercial products began to emerge in the early 2000s, from lightweight nanotechnology-enabled automobile bumpers to deep-penetrating therapeutic cosmetics. Almost everyone in technically advanced countries has used a product incorporating nanotechnology. However, almost no one knew that nanotechnology enabled the product. That remains true today.

It is fair to argue that nanotechnology research and application is a multidisciplinary research category that brought with it unprecedented optimism and serious concerns. It is interesting to note that concerns surrounding the health-related issues of nanoparticles were on the radar screen early on, long before significant data surfaced on their effectiveness as a nanoweapon.

I provided this background to make three points clear.

1. Nanotechnology is real.

2. Nanotechnology is a multidiscipline research category, with serious research beginning in 2000.

3. Even from its modest beginning, nations recognized nanotechnology's commercial and military value.

There are only two fundamental approaches to building nanotechnology. The first is "bottom-up," often referring to the method of using atoms and molecules to assemble a nanostructure. In some cases, the molecular components assemble themselves chemically by using the principles of molecular recognition (the chemical bonding of two or more molecules) to build nanomaterials and devices. The bottom-up approach implies atomic control. The second approach is "top-down," in which nanoscale objects result from larger entities, typically by removing material without atomic control during their construction. We will briefly explore each below.

1. Bottom-up approaches—This includes methods that seek to arrange smaller components into larger complex assemblies. We already discussed one popular method above, using the tip of a scanning tunneling microscope to touch the atoms and move them. More often, the atomic force microscope (AFM) is used by technologists to move atoms. The difference between the scanning tunneling microscope and the atomic force microscope is that the AFM gently touches the surface to derive an image and the STM does not. In the example above, Eigler arranged Xenon atoms on a nickel surface to spell the acronym IBM. A similar method, dip pen nanolithography (DPN), is used to create patterns directly on a range of substances with a variety of "inks." In many ways, DPN is similar to using a quill pen to write on paper. In this case, the technologist coats the DPN tip with a chemical compound or mixture acting as ink and brings it into contact with a surface acting as paper. One of the most important applications of DPN is to deposit biological materials to fabricate biosensors.

Synthetic chemistry is another bottom-up method to form molecules with a specific structure. In synthetic chemistry, specific chemicals interact to obtain a product. This technique relies on molecular self-assembly, where the molecules automatically arrange themselves into a useful structure. One important application of synthetic chemistry is to form nanomaterials, such as nanoparticles.

2. Top-down approaches—These approaches start with an element at the macro-micro level and remove material to produce an element at the nano level. You can envision this by considering the way an artist sculpts a figure from stone by using a chisel to remove material. Some of the best examples of this are in the integrated circuit industry. For example, technologists use DPN to deposit a resist, a type of chemical mask that is definable by light or electrons. Once the resist defines the circuit pattern, an etching process follows to remove material and result in a nano-level electronic transistor. This top-down fabrication methodology is termed "nanolithography." Nanolithography is also a technique used to create nanoelectromechanical systems, as demonstrated by IBM in 2000. Nano-electromechanical systems integrate mechanical and electrical functionality at the nanoscale.

Although all nanotechnology fabrication will fall under the bottom-up or top-down approaches, the multidisciplinary approach to nanotechnology has significantly expanded the way in which nanotechnology-based products are built. For example, "nano-biotechnology," which is a blanket term that refers to the blend of biology and nanotechnology, uses biology-inspired approaches to develop nanotechnology. One recent approach in nanobiotechnology is to use microorganisms to synthesize nanoparticles. The list of new processes is growing rapidly.

This chapter would not be complete without discussing the unusual quantum mechanical effects that occur at the nanoscale. Quantum mechanics is the physics that explains the behavior of mass and energy at the level of atoms and subatomic particles, the nanoscale. In general, the bulk properties of a material are significantly different at the nanoscale. Quantum mechanics dictates the properties of nanoscale materials, including melting point, fluorescence, electrical conductivity, magnetic permeability, and chemical reactivity. In addition, a specific property becomes size dependent at the nanoscale. For example, a large nanoscale material will generally be more conductive than a small nanoscale material, even when the material is the same. This occurs because the "mean free path" of electrons is greater in a larger nanomaterial than it is in a smaller nanomaterial. Even a highly conductive metal will become less conductive, more resistive, as its size enters the nanoscale. Let us examine the properties of gold at the nanoscale to understand more about the quantum mechanical effects at the nanoscale level.

If we had two vials of gold, each holding specific diameter nanoparticles of gold in a colloidal suspension, the visible color (fluorescence) of one might appear red and the other blue. The red color emits from the vial with gold nanoparticles in the range of 30 nanometers in diameter. As the gold nanoparticles increase in size to about 90 nanometers, the color changes to blue. Nothing is yellow, the color of the bulk material. The reason for this is that the size of the nanoparticle will confine the motion of the gold's electrons and, as a result, interact with light differently.

The quantum effects at the nanoscale are critical. They enable a wide range of applications. There is even a research category, quantum nanoscience, devoted to understanding the quantum mechanical effects at the nanoscale. Quantum nanoscience seeks to develop new types of nanodevices and nanoscale materials, based on their quantum mechanical behavior.

This chapter's objective is to provide a foundation to understand the fundamental techniques that nanotechnology researchers employ to build nanotechnology structures and the quantum mechanical challenges they face. We have learned that nanotechnology is not new. Indeed, nature has been using nano processes to build nanostructures for billions of years. In 2000 humans began to fabricate nanostructures, and it emerged as a research field. Since then, nanotechnology researchers have been making remarkable strides in fabricating products. These products influence the lives of most of us, even though we are unaware that nanotechnology enables them. Unfortunately, looming among the remarkable nanotechnology-based products, and typically hidden from public view, are nanoweapons.

3 I Come in Peace

> If technology is the engine of change, then nanotechnology is the fuel
> for humanity's future. Like any fuel, we must understand its usable
> capacity for doing work and apply that knowledge toward address-
> ing humanity's needs.
>
> —NATASHA VITA-MORE

The commercial applications of nanotechnology are ubiquitous. Almost every high-tech company is using nanotechnology to fabricate products or incorporating nanoscale structures in their products. This raises a serious question, When is a product categorized as nanotechnology?

Defining a nanotechnology product is challenging. For example, let's consider the definition for a nanomaterial, European Commission versus International Council of Chemical Association:

EC: Nanomaterials contain 50 percent or more particles with nanoscale features.

ICCA: Nanomaterial contain a certain percentage weight of nanoparticles with nanoscale features.

Although the definitions are similar, they are also significantly different. Both are difficult to quantify. Currently there is no internationally recognized definition of a nanotechnology product.

Categorizing nanotechnology products turns out to be equally difficult. The 2012 International Symposium on Assessing the Economic Impact of Nanotechnology delineated six distinct nanotechnology categories: transportation and aerospace, nanomedicine, electronics, energy, advanced materials, and food and food packaging.

Organizing a symposium by the above categories is appropriate for technologists, but it makes it difficult to ascertain which technologies within a category have a consumer, industrial, medical, or military application. For example, the category of advanced materials is likely to include nanotechnology with multiple applications, from consumer to military. In the interest of clarity, I am going to categorize nanotechnology products into four categories: consumer, industrial, medical, and military.

In general, the products in each category will be different, but there are likely to be overlaps in technology. This happens in the real world with numerous technologies. For example, the technology to make rifle barrels is in many cases identical for consumer hunting rifles as for military assault rifles. The steel used in industrial bridge construction can be identical to the steel used in military armor plating. The main point is that nanotechnologies are enablers. Although the nanotechnology may be identical, the applications may vary widely.

To be consistent, I will define a product to be nanotechnology if it is man-made and physically fits one of the following criteria:

Has at least one nanoscale dimension, 1–100 nanometers, or

Is constructed atom by atom or nanoscale layer by layer

Under this definition some sunscreens would be nanotechnology products, since they are man-made and contain nanoparticles. However, the abalone shell would not be a nanotechnology product, since Mother Nature makes it, even though it is constructed layer by layer, with each layer being 50 to 200 nanometers thick.

Another important concept to understand is the revolutionary nature of nanotechnology. Such products offer new properties related to the size of the nanostructures they contain. Reducing the size of a material can change its properties. For example, when I began my career in integrated circuit development and manufacture, device dimensions were in microns (a unit of length equal to one-millionth of a meter). At those dimensions, we did not find quantum mechanical effects, but we still encountered difficulties related to size, such as electromigration. Electromigration is the movement of metal ions in a conductor due to the momentum transfer between conducting electrons and the metal atoms. The electron current acted like a river that pushed the metal atoms downstream to the point where a gap appeared in the metal conductor. This resulted in open circuit failures. Electricians wiring a house, using large copper wires, do not experience this phenomenon. It surfaced in integrated circuits because the interconnecting metallurgy was in the micron range. Its emergence was unexpected. While I was employed by IBM, we worked day and night to develop a new interconnection metallurgy that was resistant to electromigration. I relate this to point out that new technology often presents new challenges related to size. Nanotechnology presents numerous and exceptional new challenges related to size, including quantum mechanical effects and surface area effects. In general, the surface area to volume of nanoscale structures has pronounced effects on how the nanoscale structures physically interact with their environment. Because of the nanoscale surface area and quantum mechanical effects, it is fair to assert that nanoscale structures exhibit revolutionary properties. In fact, many consider nanotechnology to be ushering in the sixth technology revolution.

Let's start our classification of products with some generalities. From approximately 2000 through 2004, nanotechnology products relied on the nanoscale properties of materials. These first-generation products incorporate passive nanostructures. Examples include coatings, nanostructures, and nanoparticles. In 2005 we began to see the emergence of active nanostructures. Examples include three-dimensional transistors and associated electronic circuits. By 2010 nanosystems began to emerge. Examples include Zettl's machines, like the nanoradio and the nanoelectromechanical systems. From 2015 through 2025, we expect to see the emergence of molecular nanosystems, where each individual molecule, with a specific structure, plays a specific role similar to a biological system. Applications for molecular nanosystems include nanoscale genetic therapies and cell aging therapies.

These product examples are not pipe dreams. Some estimates place the worldwide market for nanotechnology products at $1 trillion in 2015. The same forecasters expect this to grow to $3 trillion in 2020. If it continues to grow at the present rate, I estimate the worldwide market for nanotechnology products will be $6 trillion in 2025. An estimate of the worldwide economy in 2020 is $90 trillion, which suggests nanotechnology products will account for almost 7 percent of the worldwide economy. This excludes nanoweapons as products, which would obviously increase the worldwide market for nanotechnology products. One word of caution regarding these forecasts: there are differences of opinion. However, almost all agree that nanotechnology-based products are becoming mainstream and a noticeable portion of the gross domestic product of the larger players, such as the United States and Japan. Given the stakes, economically and militarily, it is no surprise that the number of worldwide nanotechnology patent applications grew 34.5 percent between 2000 (1,197) and 2008 (12,776).

In the first three categories (consumer, industrial, medical), I list nanotechnology products that are typical of the category. The list is not inclusive of all products. New nanotechnology products are hitting the market weekly, making it impossible to create a comprehensive list. I will introduce the category of "military" in this chapter and discuss it fully in the next.

1. Consumer Nanotechnology Products

In 2005 the Woodrow Wilson International Center for Scholars and the Project on Emerging Nanotechnologies established the Nanotechnology Consumer Products Inventory. The CPI is not comprehensive, since new nanotechnology consumer products appear weekly. It does, however, list over 1,800 nanotechnology-based consumer products, along with the company that produces the product. The CPI has eight major categories: Appliances, Automotive, Cross Cutting, Electronics and Computers, Food and Beverage, Goods for Children, Health and Fitness, Home and Garden, it also lists the country of origin, the product's nanomaterial, and "How much we know." Often the CPI will omit information when the information is unknown. For example, they may omit the product's nanomaterial. Each product they list includes a link to a page with more information. The CPI was updated in 2013, based on the work of research team Marina E. Vance, Todd Kuiken, Eric P. Vejerano, Sean P. McGinnis, Michael F. Hochella Jr., David Rejeski, and Matthew S. Hull, which interviewed sixty-eight nanotechnology experts. After revising the CPI in 2013, the research team published an article in 2015, "Nanotechnology in the Real World: Redeveloping the Nanomaterial Consumer Products Inventory." I will briefly summarize the results of their work:

> CPI was updated to include 1,814 consumer products from 622 companies in 32 countries.

> The Health and Fitness category contains the most products (762, or 42 percent of the total).

> Silver is the most frequently used nanomaterial (435 products, or 24 percent).

> A full 49 percent of the products (889) included in the CPI do not provide the composition of the nanomaterial used in them.

> About 29 percent of the CPI (528 products) contain nanomaterials suspended in a variety of liquid media and dermal contact is the most likely exposure scenario from their use.

> The majority (1,288 products, or 71 percent) of the products do not present enough supporting information to corroborate the claim that nanomaterials are used.

Let us examine a sample of products that are likely to influence how you work and play. For example, if you peruse the category Health and Fitness, you will find a number of subcategories. Among them is sporting goods. In this subcategory are 131 nanotechnology-based products, including golf clubs, golf balls, skis, snowboards, bowling balls, tennis rackets, badminton rackets, bicycles, and even swimsuits. Manufacturers of such products claim the products are significantly better than a competitor's product. Typical claims range from "stronger" to "lighter" and everything in between. The list of sporting goods products makes it likely that many sport enthusiasts use nanotechnology-based products. I suspect few know that nanotechnology enables their products.

Another category that finds high usage among consumers is Electronics and Computers. For example, if you recently purchased a new computer with an Intel® Core™ M processor, you are using nanotechnology. Intel states on their website:

> Thin, light, quiet, and versatile. The new 6th Generation Intel® Core™ processors built on 14 nm manufacturing process technology deliver great mobile performance, blazing-fast responsiveness, and amazing battery life. This feature-packed processor with built-in security is ready to take your productivity, creativity, and entertainment to the next level. And, by enabling the exciting Windows® 10 features, 6th Generation Intel® Core™ m3/m5/m7 processors empower you to unleash your imagination and explore the possibilities.

Although we should take manufactures' sales collateral with the proverbial grain of salt, I judge that many of the properties claimed by the manufacturers of nanotechnology-based products have a strong element of truth associated with them. Nanotechnology is a game changer, and it is likely that some manufacturers based their claims on empirical data.

The most popular category is Health and Fitness, accounting for half of all nanotechnology-based products. The Health and Fitness subcategory of cosmetics appears extremely popular and includes sunscreens, skin moisturizers, antiwrinkle products, hair care, and skin cleansers.

Currently manufacturers are not required to disclose that their products are based on nanotechnology. For example, did you know that some plastic beer bottles use clay nanoparticles to keep oxygen out and the carbon dioxide in, which keeps the beer from spoiling? In addition, some manufacturers will use the word "nano" improperly. For example, Beretta, the world's oldest gun manufacturer, labels its smallest automatic pistol as nano.

One thing is certain, the world around you has an abundance of consumer nanotechnology-based products, and the number is growing daily. You may be wearing nanotechnology-based socks, playing golf with nanotechnology-based golf clubs, while using a nanotechnology-based sunscreen to protect your skin from sunburn. At this point, that may seem fine, even desirable. However, large concerns are looming. Some nanotechnology-based products may be toxic. The reason is that many contain nanoparticles, including carbon nanotubes, nanosilver, nanometal oxides, and carbon fullerenes.

The U.S. Environmental Protection Agency and the Food and Drug Administration are beginning to grapple with the issue, debating whether nanotechnology-based products merit special government regulation. In Europe, the Health and Consumer Protection Directorate is following suit. I mentioned earlier that there is no regulation on nanotechnology-based products or nanoparticles. It is fair to raise the question, If our bodies can absorb the product's nanomaterial, as is the case with cosmetics, should it be classified as a drug and regulated accordingly?

The environmental impact and the potential hazard that nanoparticles pose to living organisms is mostly unknown. We do know that the National Nanotechnology Initiative (NNI) funding for risk research between 2000 and 2010 was only 4 percent of the total NNI budget. The 2016 NNI budget proposes that 7 percent be focused on Environment, Health, and Safety (EHS). We will also see in the next chapter that nanoparticles play an important role in nanoweapons. This suggests that more research regarding the potential hazard they pose in consumer products warrants attention.

2. Industrial Nanotechnology Products

Our primary focus in this section will be the use of nanotechnology in construction and manufacturing.

Construction is already benefiting from the use of nanotechnology in the fields of concrete and steel. Concrete and steel form the foundation of modern society from single-family dwellings to the largest urban skyscrapers.

The use of concrete-like materials dates back to 6500 BC by the Nabataea traders or Bedouins in the regions of southern Syria and northern Jordan. The ancient Greeks used concrete-like materials in the period 1400 to 1200 BC. The ancient Romans used concrete extensively from 300 BC to 476 AD, including concrete that would set underwater. There is wide agreement that concrete is the most ubiquitous material in the world.

Nanotechnology is finding wide application to improve concrete. For example, carbon nanotubes (CNTs) are commonly used to strengthen concrete. Adding titanium dioxide nanoparticles to concrete enables it to break down dirt and pollution, which allows it to wash off with rainwater. This makes buildings self-cleaning, significantly reducing maintenance costs.

Steel has been used in construction since the 1870s. However, steel can suffer fatigue, a structural failure that can shorten the useful life of structures, especially bridges. The fatigue issue requires limiting the allowable stresses and implementing frequent inspections. But this does not eliminate the issue. In fact, according to a forecast by Utah State University graduate student Wesley Cook, "Based on the data extrapolation and 95 percent confidence interval, the estimated average annual bridge collapse rate in the United States is between 87 and 222 with an expected value of 128." In Minnesota, not far from my home, the Interstate Highway 35W Bridge in Minneapolis collapsed without warning on August 1, 2007. For me, this was a shocking revelation, not another statistic. It collapsed during the evening rush hour, killing 13 people and injuring 145 others. Although bridges collapse at about the rate of one every three days in the United States, this bridge was within a short driving distance of my home and used by family, friends, and employees. Although this collapse was due to a design flaw, according to the National Transportation Safety Board, some consultants suspect that steel metal fatigue also played a role.

According to structural engineer Surinder Mann, "Stress risers are responsible for initiating cracks from which fatigue failure results and research has shown that the addition of copper nanoparticles reduces the surface unevenness of steel which then limits the number of stress risers and hence fatigue cracking." In general, there are two ways to improve the strength and fatigue resistance of steel.

1. Eliminate defects by reducing the size of each crystal, within the bulk material, to the point that there is literally no room for defects

2. Introduce a large density of defects that act as an obstacle to block the motion of dislocations.

To achieve these desirable properties, the steel industry is turning to nanotechnology. The addition of various nanoparticles, such as copper, carbon nanotubes, vanadium, molybdenum, magnesium, and calcium, is making steel stronger and fatigue-resistant. The challenge is to manufacture large components of bulk nanotechnology-enhanced steel at a reasonable cost. To date, the steel industry has developed over a dozen processing strategies, and companies in the United States, Japan, and elsewhere are filing patents to protect their developments. Applications in construction are numerous, ranging from steel cables to bolts. Especially interesting are the nanotechnology coatings in development to enhance both the strength and corrosion resistance of steel. In 2015 MIT Technology Review reported, "An inexpensive new process can increase the strength of metals such as steel by as much as 10 times and make them much more resistant to corrosion." The new nanocoating is the result of "an advanced form of electroplating" that offers "precise control over the structure of metals." It is evident that the use of nanotechnology in concrete and steel offers the potential to revolutionize construction.

Let us now discuss the use of nanotechnology in manufacturing. All nanomanufacturing relies on top-down or bottom-up approaches. Within those approaches, many new methodologies are enabling nanomanufacturing. According to NNI, examples include:

Chemical vapor deposition—a chemical reaction that produces pure, high-performance films.

Molecular beam epitaxy—a method for depositing precise thin films.

Atomic layer epitaxy—a method that deposits one-atom-thick layers on a surface.

Nanoimprint lithography—in this process nanoscale features result by "stamping" or "printing" them onto a surface.

Roll-to-roll processing—a high-volume process that produces nanoscale devices on a roll of ultrathin metal or plastic.

Self-assembly—a process in which components come together to form an ordered structure on their own.

Let me give you an example of top-down nanomanufacturing from personal experience. During my career in the integrated circuit industry, we manufactured integrated circuits with submicron geometries, approaching the nanoscale using electron beam lithography. In essence, we would first deposit a thin electron-sensitive resist material on a silicon integrated circuit substrate, which is the structure that embodies the individual silicon chips before they are diced (cut apart). We then would expose the resist by scanning an electron beam in a desired pattern across a resist-coated substrate. The equipment to do this is commercially available. This is similar to photolithography but uses electrons instead of photons. It also eliminates the need for a conventional photomask. The electron beam lithography is a direct write methodology, meaning the electron pattern was computer-controlled to result in the desired device feature. The elimination of a photomask removes one process step, which offers the potential to lower costs and eliminate defects. Electron beam lithography enables nanotechnology manufacturing because electrons are extremely small elementary particles with a negative charge. Their negative charge allows direct write control. The size of electrons enables nanoscale features. Regarding the size of electrons, if you compare the hydrogen atom nucleus to a baseball stadium, an electron would be about the size of a bee circling the stadium. Today electron beam lithography is in wide use to create features as small as 10 nanometers.

The example above demonstrates nanomanufacturing of integrated circuit devices. Three important benefits result from shrinking integrated circuit devices into the nanoscale. First, nanoscale devices are faster and use less power. Second, they enable the manufacturer to increase the number of integrated circuits that a given substrate holds, which reduces manufacturing cost. Lastly, new properties, only present at the nanoscale, enable increased integrated circuit functionality.

Integrated circuits incorporating nanoscale devices enable numerous applications. For example, many within the nanotechnology community agree that the nanoweapon laser shown in figure 1 is only possible via nanoscale components. The U.S. government is not going to comment on its construction. There is no mention of it on the NNI website, which appears intentional. In general, the NNI website provides little information on the use of nanotechnology in warfare. After searching the NNI website, I found several articles focused on soliciting basic research projects that support the Defense Threat Reduction Agency mission to safe-guard America and its allies from weapons of mass destruction.

According to NNI, "NNI agencies are investing heavily in nano-manufacturing R&D and infrastructure. Over 90 NNI-funded centers and user facilities across the country provide researchers with the facilities, equipment, and trained staff to develop nanotechnology applications and associated manufacturing processes." NNI takes great pride in its role relating to the development of nanotechnology products and manufacturing. It devotes an entire section on the website to benefits and applications, and the list is impressive. However, it does not mention a single military or weapon application. Almost certainly, the DoD held a conference similar to the 2012 International Symposium on Assessing the Economic Impact of Nanotechnology to discuss military nanotechnology research, nanomanufacturing, and nanoweapon applications. However, that conference only included defense contractors engaged in nanoweapons research, manufacturing, and applications, along with associated DoD leaders. I attended similar conferences regarding Honeywell's work on DoD programs. Those conferences were secret, and their proceedings were withheld from the public.

3. Medical Nanotechnology

The medical application of nanotechnology in medicine is termed "nanomedicine." According to the editor in chief of the *Journal of Human Reproductive Sciences*, Mallanagouda Patil, nanomedicine is "the science and technology of diagnosing, treating, and preventing disease and traumatic injury, of relieving pain, and of preserving and improving human health, using nanoscale structured materials, biotechnology, and genetic engineering, and eventually complex machine systems and nanorobots."

Current applications in nanomedicine include nanomaterials, biological devices, and nanoelectronic biosensors. Future applications are likely to include biological machines and nanorobotics. One of the most important problems that nanomedicine is addressing is the toxicity and environmental impact of nanoscale materials, especially nanoparticles. Unfortunately, man-made pollution produces nanoparticles, and some nanoparticles pose a health hazard.

Nanomaterials are particularly useful in nanomedicine, since their size is in the same range as biological molecules. Nano-technologists are working to bond nanomaterials with biological molecules to construct "drug carriers." This is leading to an entirely new class of nanomedicines, "smart drugs." Smart drugs are analogous to smart bombs. Experts at the European Medicines Agency assert, "The majority of current commercial applications of nanotechnology to medicine are geared towards drug delivery to enable new modes of action, as well as better targeting and bio-availability of existing medicinal substances. Novel applications of nanotechnology include nanostructure scaffolds for tissue replacement, nanostructures that allow transport across biological barriers, remote control of nanoprobes, integrated implantable sensory nanoelectronic systems, and multifunctional chemical structures for drug delivery and targeting of disease."

Many bacteria and viruses are nanoscale in size. It is logical to use nanoscale treatments, nanomedicine, to fight bacteria and viruses on their own level. The ancient Greeks used silver to promote healing of wounds. Today physicians use bandages impregnated with silver nanoparticles to treat burn victims. The silver nanoparticles are more active than the bulk material due to their increased surface to volume ratio and their ability to easily penetrate the skin. In addition, unlike antibiotics, bacteria and viruses appear unable to build a resistance to silver nanoparticles. As a result, burn victims heal faster, and their bandages do not require frequent and painful changes.

One of the most exciting uses of nanomedicine is in diagnosing and treating cancer. Researchers Edward Kai-Hua Chow and Dean Ho published a 2013 paper in *Science Translational Medicine* on the efficacy of using nanomedicine: "Nanotechnology-based chemotherapeutics and imaging agents represent a new era of 'cancer nanomedicine' working to deliver versatile payloads with favorable pharmacokinetics and capitalize on molecular and cellular targeting for enhanced specificity, efficacy, and safety."

There are numerous other areas of nanomedicine, including tissue engineering and nanosurgery. Nanomedicine is moving into the mainstream. According to the Kidlington Centre, the nanomedicine "market is on a steady growth, and with a compound annual growth rate (CAGR) of 12.5 percent from 2011, the market size will reach to $130.9 billion by 2016."

Closely related to nanomedicine is nanoelectronics in medicine. Conventional electronics, including integrated circuits, find ubiquitous use in medical monitoring of vital signals, pacemakers, drug delivery, and limb stimulation. Nanoelectronics will accelerate implanted electronics to facilitate both diagnostics and treatment of numerous illnesses, with cancer heading the list.

There is an immense body of literature discussing nanomedicine. The focus of this section is to raise the awareness of the use of nanotechnology in medicine. However, let us address the most important question, What long-term benefits should we expect from nanomedicine?

According to Robert A. Freitas Jr., a senior research fellow at the Institute for Molecular Manufacturing in Palo Alto, California, "The ultimate goal of nanomedicine is to perform nanorobotic therapeutic procedures on specified individual cells comprising the human body." Freitas reasons, "Most human diseases involve a molecular malfunction at the cellular level." He envisions using nanobots about the size of blood cells that would traverse the human body using its vascular system to treat individual cells via genetic therapy by removing the damaged genetic material from the cell and inserting newly manufactured genetic material that would restore normal cell function. Freitas's paper is "the first theoretical scaling analysis and mission design for a cell repair nanorobot." To some, Freitas's vision may come off as science fantasy. In my opinion, it is not. I see it as a futurist's perspective on how nanomedicine may evolve to treat disease at the cellular level. Given the enormous strides in nanotechnology in just the past decade, I judge we may soon see Freitas-type nonorobots, which he calls "chromallocytes," emerge.

4. Military

Sections 1–3 make a strong case that the consumer, industrial, and medical uses of nanotechnology are becoming ubiquitous, as governments and companies take advantage of nanotechnology's unique properties. It is also apparent that nanotechnology has the potential to enable a quantum leap toward a brighter future for humanity, from curing inoperable cancer to providing ultra-fast nanoelectronic computers. However, nanotechnologies and nanomanufacturing also embody the potential for disaster, including human extinction. Technology is ethically neutral, but applications give rise to ethical issues. For example, the same nanotechnology that finds application in making concrete stronger, thus enabling taller skyscrapers, can also find application in weapons, like military bunkers. Indeed, in the next chapter, we will see that many consumer, industrial, and medical applications can form the basis of nanoweapons.

Our history as a species gives testimony to two indisputable facts. First, we engage in wars. Wars have plagued humanity since ancient times. Second, each war gives rise to weapons with greater destructive capability. At the end of World War II, we entered the nuclear age. Nuclear weapons were our first weapons of mass destruction. Even today, there are sufficient nuclear weapons to wipe out humanity twice over. We have not used nuclear weapons in any conflict since 1945. The reason for their lack of use does not stem from some well-thought-out ethical foundation. It is based on the doctrine of mutually assured destruction. This has kept the use of nuclear weapons at bay. It is simple survival. If we use nuclear weapons, our adversary will also use nuclear weapons. In effect, we ensure our mutual destruction.

Similar thinking went into banning the use of biological weapons. In a 1969 press conference, President Richard Nixon stated,

"Biological weapons have massive, unpredictable, and potentially uncontrollable consequences." He added, "They may produce global epidemics and impair the health of future generations." In 1972 Nixon submitted the Biological Weapons Convention to the Senate. By laying this groundwork, the Prohibition of the Development, Production, and Stockpiling of Bacteriological (Biological) and Toxin Weapons and on Their Destruction proceeded to become a multilateral disarmament treaty in 1975.

In my view, the year 2000 marked our entry into the nanotechnology age. Nanoweapons, similar to nuclear and biological weapons, have the potential to render humanity extinct. Will we recognize the danger that nanoweapons pose to the survival of humanity prior to using them in armed conflict, or will it require equivalent Hiroshima and Nagasaki events to make their massive destructive power visible?

4 The Wolf in Sheep's Clothing

> If we continue to develop our technology without wisdom or prudence, our servant may prove to be our executioner.
>
> —GEN. OMAR N. BRADLEY

This chapter discusses how nanotechnology in consumer, industrial, and medical applications can be transformed into nanoweapons. For example, the same nanoscale laser that may find nanosurgical applications to remove a cancerous tumor cells can form the basis of a mini-nuke nanoweapon. It also discusses nanoweapons specific to military applications that have no commercial counterpart.

To bring order to our discussion, I have developed five categories.

1. Nanoweapons pervasive in all branches of the U.S. military

2. U.S. Navy nanoweapons

3. U.S. Army nanoweapons

4. U.S. Air Force nanoweapons

5. Nanoweapon thrusts of other nations

My research indicates that the United States leads in nanoweapons, but the full scope of that lead is closely guarded. It is likely that all nanoweapons are classified above Top Secret, with the "need to know" rule rigorously enforced. My point is that it is possible that no one in the entire United States knows the full details about nanoweapons in all categories, including the president of the United States.

With the above understanding, let us begin.

1. Nanoweapons Pervasive in All Branches of the U.S. Military

In 2009 the DoD published its "Defense Nanotechnology Research and Development Program." The report asserts,

> The DoD nanotechnology program is based on coordinated planning and federated execution among the military departments and agencies (components) (e.g., the Defense Advanced Research Projects Agency [DARPA] and the Defense Threat Reduction Agency [DTRA]). Nanotechnology currently represents a scientifically and technologically advanced research theme that has proven and expected value toward enhancing defense capabilities. This report was compiled by a team of personnel from all the DoD Components involved in nanotechnology research and technology and associated with the National Nanotechnology Initiative (NNI). Detailed technical and programmatic reviews of nanotechnology efforts were collected and collated to respond with current, planned, and expected projects and were augmented with studies to summarize global research efforts, transition progress, and industrial base status and requirements.

Although dated, this report provides important historical insight into the DoD's nanoweapons research and development. Much of what is contained in this chapter result from that report. In places where the report does not sufficiently address nanoweapons, we will use a wide spectrum of other literature to paint a complete picture.

Without doubt, the number one pervasive nanoweapons are nanoelectronic integrated circuits. All branches of the U.S. military, as well as government agencies, use computers. As mentioned earlier, Intel's recent generation of processors employ nanotechnology. Therefore, as the U.S. military and specific government agencies procure new computers, they are deploying nanoweapons. This is an obvious nanoweapons application, but worth noting as indicative of commercial technology finding nanoweapon applications. Less obvious is the use of radiation-hardened nano-electronic integrated circuits. These find wide application in any electronic systems likely to encounter the high levels of radiation, such as during a nuclear detonation. Weapon systems include missiles, surveillance satellites, communication satellites, and nuclear weapons, to name a few. In general, radiation-hardened integrated circuit development and manufacturing is similar to commercial integrated circuit development and manufacturing, differing in two ways, process and design.

The first and most important is process. Radiation-hardened integrated circuits processing employs insulating substrates, such as silicon on insulator (soi), instead of the usual silicon substrates. This prevents the ionizing radiation from causing current leakage to the substrate. Commercial integrated circuits can withstand between 50 and 100 "gray," where a gray is the absorption of one joule of radiation energy per one kilogram of matter. A radiation-hardened integrated circuit on soi can withstand at least ten times that amount and, with proper design, even more. The design of radiation-hardened integrated circuits is significantly different from that of their commercial counterparts and depends on the circuit's application. The actual design methods are classified.

The second pervasive nanoweapons are nanoparticles. These are finding applications in nanomedicine, nano coatings, nano-enhanced materials, and nanotechnology-based explosives. In the last chapter, we discussed the use of nanoparticles in nanomedicine. Every branch of the military is making use of nanomedicine as it becomes available. Nano coatings are finding widespread application, especially in the U.S. Navy. Nano-enhanced materials, such as nanoengineered metals, brings new meaning to the phrase "stronger than steel." Nanotechnology-based explosives will eventually replace conventional explosives due to the increase in their destructive force. For example, the DoD is using nanoaluminum to create ultrahigh burn rate chemical explosives, with a ten times greater explosive punch than conventional explosives. This enables smaller payloads, about a tenth of the amount, with an equal or greater destructive power. Imagine, for example, a mortar projectile that is capable of taking out a fortified enemy command installation.

The third pervasive nanoweapons are nanosensors. The military uses numerous sensors in every branch. Nanosensors offer unparalleled opportunities to interact (i.e., sense) at the molecular level. This makes them extremely effective as biosensors and chemical sensors, where the requirement is to detect low concentrations with high specificity. Nanobiosensors and nanochemical sensors offer significant potential in the detection of explosives, biological warfare agents, and chemicals. The realities of modern warfare are driving the DoD to develop "lab-on-a-chip" devices (i.e., integrated circuits that incorporate sensors). The goal is to manufacture reliable, inexpensive devices with technology similar to integrated circuits. The low cost, weight, and size efficiency of nanosensors make them especially attractive. Ideally, each soldier could carry with them a lab-on-a-chip that monitors their health and environment. Is this a nanotechnology pipe dream? Not according to a 2007 MIT *Technology Review* article, "Nanosensors in Space," by Brittany Sause:

> NASA scientists built the nanosensors by coating carbon nanotubes with different polymers that react with various chemicals, or by doping the carbon nanotubes and nanowires with different catalytic metal particles that act as the sensing material. There are 32 sensing channels on a chip; depending on the chemical to be detected, various nanostructure materials, carbon nanotubes, or metal-oxide nanowires— coated or uncoated—are put in each channel. When a small amount of the targeted chemical touches the sensing materials, it triggers a reaction that causes the electric current flowing through the sensor to increase or decrease. The different responses will form a pattern, which the sensor can use to identify a gas.

The entire sensor package is about 5.1 by 6.4 by 2.5 centimeters.

THE WOLF IN SHEEP'S CLOTHING

(Note: 2.5 centimeters is approximately equal to one inch.) In addition, it is wireless and can transmit sensor data electronically. The technology can be tuned to detect various agents by changing the nanostructures in the various channels. Given that the MIT article is in the public domain suggests even more advanced nanosensors are under development.

The fourth and final pervasive nanoweapons are nanorobotics. Today's military makes use of robotics in every aspect of warfare, from surveillance to offensive operations. Nanorobotics will extend the use of robotics to a wider range of critical applications.

Please note that pervasive nanoweapons have a strong tie to similar nanotechnologies used in consumer, industrial, and medical applications. This is an important point. Nanotechnologies are enablers. The specific application dictates the category. If nanoparticle enhanced steel is a structural element in a building, we view it as an industrial application. If the same steel finds application in tank armor, we view it as a military application.

Other nanoweapons are not as pervasive as nanoelectronics, nanoparticles, nanosensors, and nanorobotics. Therefore, we will cover them in the context of the military branches.

One point deserves mention in this section. Governments, working with the private sector, generally find it easier and less costly to construct a nanoweapons manufacturing facility versus a nuclear weapons facility. For example, the equipment used by radiation hardened integrated circuit manufacturers, like Honeywell, is commercially available. Although the integrated circuit companies using nanolithography must still invest in process technology to optimize the manufacture of nanoscale features, their starting point is commercially available equipment and process materials. Another example is the manufacture of nanoparticles, which play an important role in consumer and industrial products, as well as nanoweapons. In 2014 the University of Massachusetts Amherst announced "a breakthrough technique for creating water-soluble nano-modules and controlling molecular assembly of nanoparticles over multiple length scales." They state that their new technology will reduce the time that nanotechnology companies spend to develop nanoscale materials using traditional methods. In many cases, it is possible to establish nanoweapons manufacturing in a relatively small facility, like a small office building, compared to a nuclear weapons manufacturing facility, and at a fraction of the cost. In addition, detection of nanoweapons manufacture is extremely difficult compared with detection of nuclear weapons manufacture, which gives off a telltale radioactive signature. Lastly, international treaties do not regulate nanoweapons. They do regulate nuclear weapons.

Before we proceed to the next section, let me add perspective. This chapter represents a snapshot in time. With the passage of time, other nanoweapons may become pervasive. Like all military systems, nanoweapons will evolve as nanotechnology evolves. Specific nanoweapons may become obsolete with new nanoweapons taking their place. Entirely new nanoweapons may emerge as nanotechnology advances. In short, we should view any discussion regarding nanoweapons as dynamic, subject to change as nanotechnology and threat assessments evolve. With this understanding, let us continue.

2. U.S. Navy Nanoweapons

The U.S. Navy is using a wide spectrum of nanoweapons. In fact, the U.S. Naval Research Laboratory includes the Institute for Nanoscience. According to their website, "The Institute exploits the broad multidisciplinary character of the Naval Research Laboratory in order to bring together scientists with disparate training and backgrounds to attack common goals at the intersection of their respective fields at this length scale. The objective of the Institute's programs is to provide the Navy and the DoD with scientific leadership in this complex, emerging area and to identify opportunities for advances in future Defense technology."

The 2009 DoD report, "Defense Nanotechnology Research and Development Program," details the U.S. Navy's significant nanotechnology research and development (R&D). Here are the key R&D goals and challenges quoted from the report:

Fundamental Phenomena and Processes—Understand novel phenomena that are manifested at the nanometer length scale: plasmonic and coupled plasmonic/photonic behavior; phonon/electron transport; nanoscale chemical signaling; DNA-guided patterning for sub 10-nm accuracy and electron spin physics, including quantum coherent phenomena

Materials—Control the physical and chemical properties of nanoscale carbon materials and doped nanowires for electron devices, sensors, and solid-state energy conversion

Devices and Systems—Construct novel electron devices and associated circuits that are intrinsically advantageous to field-effect transistors

Manufacturing

Develop bio-enabled fabrication of electronics that bridges top-down and bottom-up approaches to achieve better than 10-nm spatial resolution

Develop electrochemical techniques for patterning metals to produce nanoscale features for antennas, sensors, meta-materials, and catalysts

Research Facilities and Instrumentation—Provide instrumentation to support research in the areas of spintronics, nanotube sorting, and three-dimensional (3-D) fabrication via the Defense University Research Instrumentation Program

Education and Societal Dimensions—The programs within the Navy nanoscience research area directly support approximately 150 graduate students and 75 postdoctoral students

The depth and breadth of the Navy's research and development of nanotechnology are formidable. As is evident, many of the thrusts are fundamental research. This is intentional. According to the DoD report, "Nanotechnology must still be considered to be in its technological and engineering infancy. Some significant applications have been noted, but these applications are viewed as just the beginning of an age of practical products based on nanomaterials and nanoprocesses. The engagement of DoD in nanotechnology research will continue to be important in ensuring the optimum direction of ongoing research efforts and the optimum leveraging of this knowledge to advance warfighter and defense-related systems capabilities." In effect, the DoD is acknowledging that it is too early to determine which nanotechnologies will be required for the next generation of nanoweapons.

Having surveyed the Navy's R&D, let us explore some of the Navy's nanoweapon applications. In chapter 1, we discussed the Navy's deployment of a laser nanoweapon on the USS *Ponce* in 2013. While this is a good example, let us survey some of the other nanoweapons the Navy is pursuing. In 2015 the Office of Naval Research published the "Naval Science & Technology Strategy." It mentions "nano" eight times, including "nanotechnology," "nanoscale materials," "coatings at the nano-meter," "nano electronics," "nanometer scale electronic devices and sensors," and "Naval Research Laboratory's Institute for Nanoscience." The report indicates that the fields of materials, electronics, and biology are of particularly high interest to the Navy.

Since 2002 the Navy has been applying nanotechnology in ship fabrication and maintenance. One of the biggest challenges the Navy faces is corrosion. The Navy uses various metals to construct ships. As a simple high school science project demonstrates, if you put two different metals in a jar filled with saltwater, you create a battery. From the Navy's viewpoint, the battery effect is undesirable, since it accelerates corrosion. To combat this issue, the Navy is using nanoceramic coatings, which provide all the protective features of conventional ceramics, but adhere to surfaces better and are far less brittle, to the point they can deform along with whatever it covers. This is critical in a battle scenario. For example, when a submarine is under depth charge attack, the hull will undergo significant deformation stresses. The nanoceramic coatings are able to flex and continue to adhere during the stresses. In fact, the first naval application using nanoceramic was on a submarine hatch made of a titanium plate bolted to the steel frame. The purpose of the hatch was to open underwater and allow the deployment of sensors and other electronics. The nanoceramic coating solved the battery issue, known as a galvanic reaction, and proved an extremely durable coating. Other Navy applications for nanoceramic coatings are to reduce friction between components, such as reducing the friction of ball valves that regulate water flow in submarines.

This is a good place to emphasize an important point. Nanoweapons are any military technology that exploits the use of nanotechnology. Thus even a naval ship coating can be a nanoweapon. Another serious problem the Navy faces is marine growth on the outer hulls of their ships. For example, barnacles can add over 60 percent drag, which wastes fuel and diminishes speed. Barnacles, sea creatures related to crabs and lobsters, are able to attach themselves to any surface under any condition, and their natural cement rivals superglue. The Navy is continually evaluating materials to stop barnacles from attaching themselves to Navy ships. In 2009 the Office of Naval Research announced research into a promising new biofriendly coating, but they did not disclose the details. However, in 2014, X. Zhao et al. published an article that provides potential insight into the Navy's new coating, "Studies on Nano-additive for the Substitution of Hazardous Chemical Substances in Antifouling Coatings for the Protection of Ship Hulls." They state: "Based on the bactericidal capacity of cuprous oxide and photochemical effect of nano-additive, environment-friendly and efficient marine antifouling paints were prepared in this study." It is conceivable that this nanotechnology-based coating is the basis of the Navy's new coating.

Lastly, let us discuss power control systems. A Navy ship, such as an aircraft carrier, supplies power throughout the ship similar to how a power company supplies power to a city. It relies on copper cables, insulators, transformers, and other bulky equipment. Unfortunately, this conventional approach results in significant weight and power loss due to Joule heating. Joule heating results when a current passes through a conductor. The heating is the result of electrons colliding with atoms of the conductor and themselves. The collisions create heat, which results in energy loss. Here is a simple example. If you get a power surge, the rubber insulation of wires leading to electrical appliances may melt. This is due to Joule heating. The Navy is turning to nanotechnology to solve these problems. According to Nanowerk, "The Office of Naval Research has awarded University at Buffalo engineers an $800,000 grant to develop narrow strips of graphene called nanoribbons that may someday revolutionize how power is controlled in ships, smartphones and other electronic devices." Graphene is a nanocarbon material that is thin, light, strong, and the best-known conductor of heat and electricity. Graphene can withstand extreme energy loads, approximately 1,000 times greater than copper. In addition, graphene can be grown, similar to growing a biological material. Finding an improved power control system is critical to the twenty-first century U.S. Navy, which appears to be driving toward electric propulsion systems. For example, the Navy recently launched its first stealth destroyer, the USS *Zumwalt*, and in 2015 began sea trials. In addition to being the Navy's largest destroyer and having a significantly reduced radar profile, about fifty times less than conventional destroyers, the USS *Zumwalt* boasts an integrated power system, which is an all-electric system powered by gas turbines. This example underscores the criticality of having an improved power control system.

There are many more applications in development, from nanowire gas sensors to nano-coated self-cleaning window glass. There are also, undoubtedly, nanoweapons buried even deeper below the cloak of secrecy.

3. U.S. Army Nanoweapons

In 2003 the U.S. Army established the Institute for Soldier Nanotechnologies (ISN). Its goal is to enhance the soldiers' protection and survivability, including "threat detection, threat neutralization, automated medical treatment, concealment, enhanced human performance, and reduced logistical footprints." The ISN lists five strategic research areas, delineated as SRAS. Under each SRA, they list themes and projects. Because the ISN's list demonstrates the depth and breadth of the Army's nanoweapon research, appendix I lists the SRAS, themes, and projects verbatim from their website.

I recognize reading ISN's list may be daunting. Unfortunately, their use of technical terms and acronyms tend to hinder understanding. To enable insight, I have defined the technical terms and acronyms in the glossary. However, even using the ISN's list and this book's glossary may still result in confusion. For this reason, I am going to present a simplified view. When most of us think about the Army, our minds conjure pictures of soldiers in battle gear, snipers, tanks, rifles, and artillery. That picture is likely to stay the same even when the ISN's nanoweapons are widely applied by the Army, but the effectiveness of every element in our picture is likely to increase dramatically. Here are enhancements the ISN is likely to provide within the next decade:

1. Soldiers wear nanomaterial that feels like cloth but is impervious to bullets, as well as biological and chemical agents.

2. Reduce the weight a solder carries in battle, often over seventy pounds during the Afghanistan and Iraq wars, to less than fifty pounds without loss of capabilities.

3. Snipers become completely invisible under a nanomaterial invisibility cloak and fire smart bullets that make a one-mile kill routine.

4. Tank armor is a nanomaterial, stronger and lighter than conventional steel, and the projectiles they fire have nano-enhanced explosives capable of leveling a large building

5. Army infantry lasers, mounted to conventional vehicles, are able to fire laser bursts of energy capable of destroying cruise missiles, artillery, rockets, and mortar rounds

6. Artillery fires "smart" projectiles with nano-enhanced explosives capable of leveling a large enemy fortification or surgically assassinating a single enemy combatant

Everything you just read may sound like science fiction. However, it is not. To make this case, we will examine each point and the nanotechnology that enables it.

Let us start with wearable nanomaterial and the rationale for its use. The body armor that infantry soldiers wear on patrol in Iraq weighs at least thirty-three pounds. While this armor offers protection against enemy rifle bullets and, to a limited extent, improvised explosive devices (IEDS), it makes agile movement difficult. In response to this issue, the Army is seeking body armor with reduced weight that would allow normal agility. Nanomaterials appear to offer a solution.

In 2012 scientists at MIT, supported by the U.S. Army Research Office, published a paper demonstrating the capability of a thin, light composite nanomaterial to stop bullets as effectively as heavyweight inch-thick body armor. However, the Army is considering numerous other solutions as well. They include passive nanomaterial, such as adding Si or TiO_2 nanoparticles embedded in epoxy matrix, SiO_2 nanoparticles in a liquid polymer that hardens on ballistic impact, and carbon nanotubes, a material lighter than plastic and one hundred times stronger than steel. In addition to passive nanomaterial, the Army is investigating smart nanomaterial, meaning the nanomaterial properties are controlled by sensors embedded in the soldier's armor. For example, if a body armor sensor detects an impact, it could trigger an electric pulse to iron nanoparticles in inert oil, which harden on electric pulse stimulation. It is also a near certainty that the next generation of body armor will include embedded sensors, capable of sensing impact, heat, chemical, and biological agents. In addition, it is likely the military would also impregnate the body armor with nanomedical agents to offer immediate first aid, such as blood clotting and antibacterial agents.

In total, the weight of the U.S. soldier's body armor and gear is over seventy pounds. As a result, soldiers are suffering musculoskeletal injuries. In fact, one-third of medical evacuations from Iraq and Afghanistan, between 2004 and 2007, were the result of musculoskeletal, connective tissue, or spinal injuries. Therefore, in addition to reductions in body armor weight, the Army seeks to reduce the total "load out" carried by the soldier, which includes a communication radio, extra batteries for the radio, water, food, flashlight, first aid kit, and ammunition. Once again, nanotechnology appears to offer solutions. Nanomaterial-enhanced batteries, such as silicon-coated carbon nanotube anodes for lithium-ion/ (Li-ion) batteries, are an attractive option. The nano-enhanced Li-ion batteries would have an energy capacity ten times that of conventional batteries and eliminate the need for extra batteries. Ultimately, nanoelectronics will enable a reduction in the size and weight of the radio and flashlight with no compromise in functionality. The Army's goal is to get the total weight a solider carries to be less than fifty pounds. In addition, it is also likely that some soldiers will be fitted with an exoskeleton to amplify their muscle strength to enable them to carry out specific missions. No doubt, the exoskeletons will use light nanomaterials stronger than steel, such as carbon nanotubes, powered with nanomaterial-enhanced batteries.

A cloak of invisibility seems to be a phrase out of *Star Trek*. However, it exists now. Scientists at the U.S. Department of Energy's Lawrence Berkeley National Laboratory and the University of California Berkeley have devised a nano-thin invisibility cloak that makes 3D objects "disappear." The ultrathin invisibility nanomaterial can conform to the shape of an object and conceal it from detection with visible light. According to the September 18, 2015, *Nanowerk News* article:

Working with brick-like blocks of gold nanoantennas, the Berkeley researchers fashioned a "skin cloak" barely 80 nanometers in thickness, that was wrapped around a three-dimensional object about the size of a few biological cells and arbitrarily shaped with multiple bumps and dents. The surface of the skin cloak was meta-engineered to reroute reflected light waves, which rendered the object invisible to optical detection.

Imagine military personnel invisible to the naked eye. Imagine combining that capability with nanomaterials able to insulate a person's body heat, similar to how polar bears' fur insulates their bodies, and to scatter infrared. In effect, a Special Forces team using skintight invisible cloaks becomes real-life ninjas able to penetrate even highly guarded installations. Conventional camouflage would become obsolete.

Today we have smart bombs. As electronics shrink to the nanoscale, it will be possible to have smart bullets. In fact, CNN reported in April 2015: "The U.S. military said this week it has made great progress in its effort to develop a self-steering bullet. In February, the smart bullets—.50-caliber projectiles equipped with optical sensors—passed their most successful round of live-fire tests to date, according to the Defense Advanced Research Projects Agency, or DARPA."

If you find this scary, imagine a bullet "with your name on it." More precisely, imagine a bullet programmed to target you via your DNA. Actually, the word "bullet" may be inappropriate. The projectiles would be similar to bullet-sized guided missiles, including nanosensors, nanoprocessors, and nano guidance systems that enable it to seek a specific enemy combatant, even when hidden behind cover. Snipers, armed with such smart bullets, could routinely make kills out to a mile or more.

Enhanced steel will likely find wide military application. In chapter 3 we discussed how adding nanoparticles and nano coatings could make the steel ten times stronger and more resistant to fatigue. The nano coating is especially interesting. Any metal can cost-effectively undergo this nano-coating process. According to Christina Lomasney, CEO of Modumetal (2016), "The process costs the same as conventional metal treatments such as galvanization." Imagine tanks half the weight, with armor stronger than today's tanks. Being lighter, they could travel faster and further with the same amount of fuel they currently consume. This is just one example. The use of metals in military applications is ubiquitous.

Nanoparticles are finding use as catalysts to increase the destructive power of conventional explosives. We discussed this earlier, but will expand further in this section. The field of nanoenergetics, which relates to manipulating the flow of energy within and between molecules, is yielding nanoweapon explosives with up to ten times the destructive capability of conventional explosives. They also offer enhanced propellants for missile applications. Researchers at Sandia Laboratories are experimenting with adding materials known as superthermites. Superthermites refer to the mixture of nanoparticles, such as iron oxide and aluminum nanoparticles, which react more rapidly than micro thermites due to the large surface to volume ratio of nanoparticles. The applications are wide-ranging, from more devastating artillery projectiles to smart bullets that not only penetrate but also explode. Typically, being grazed by a conventional bullet is not fatal. However, a smart bullet that explodes on contact is likely to be fatal to its intended target and those in close vicinity.

Mobile lasers are likely to become widely deployed by the Army. In 2009 Northrop Grumman announced the successful development of an electric laser capable of producing a 100 kilowatt ray of light, sufficient to destroy cruise missiles, artillery, rockets, and mortar rounds. Electric lasers are mountable to conventional vehicles because they require less space than chemical lasers. In this, we see science fiction, like *Star Trek* and *Flash Gordon*, becoming science fact.

Nanotechnology will transform current artillery to smart artillery, profoundly changing the role of artillery. The impact of nanotechnology on artillery may even change the role of the Army, as a land force, to a strategic force, capable of engaging air, sea, and land targets. The history of artillery in warfare dates back hundreds of years. All military scholars agree regarding its use as a primary weapon during the Civil War, World War I, and World War II. Artillery generally applies to a class of weapons, such as howitzers, cannons, mortars, and other weapons that fire a large projectile. It excludes small arms and missiles. Nanoelectronics and nanosensors have the capability to make artillery projectiles "smart," meaning they will have properties that resemble guided missiles. The projectiles will be programmable to target specific coordinates, structures, and even people. Nano-enhanced explosives will provide the smart projectiles with a new level of devastation. I project it will be possible to program their devastation to accomplish the mission while limiting collateral damage.

Although the Army is the oldest branch of the U.S. military, it makes use of high-tech equipment, including nanoweapons. The phrase "boots on the ground" may change to "nanoweapons on land."

4. U.S. Air Force Nanoweapons

The 2009 DoD report "Defense Nanotechnology Research and Development Program" provides significant insight into the Air Force's nanotechnology R&D. Quoted from the report:

> Fundamental Phenomena and Processes—Advance the technology to create the world's smallest array of imaging pixels for high-resolution, multispectral imaging systems

> Materials—Identify a better nanoparticle and develop a technique to create multilayer superconducting wire with carefully controlled particles dispersion via an amenable industrial process

> Devices and Systems—Develop 2-nm platinum-silicide (PtSi) films on silicon (Si) and image processing methods to dominate infrared (IR) camera technology

Manufacturing—Create a highly uniform array of carbon nanotubes with an electronic grade junction to a Si semiconductor substrate to develop an ultralightweight, uncooled IR detector

Environmental Health and Safety—Develop assessment studies to understand the risks associated with the physiochemical properties of size, shape, and surface chemistry

Perusing the list, an obvious question arises, What is driving the Air Force's nanotechnology needs? In simple terms, it is the increased strategic requirement for unmanned aerial vehicles (UAVs), also termed "drones." Currently about a third of all Air Force fighter planes are drones, and the need for drone missions skyrocketed during the Obama administration. Drones have proven effective in gathering intelligence and killing high-value targets. Some estimate that drones have killed over 2,500 enemy combatants, arguably some of them civilians, during the Obama administration.

Most of the nanoweapons the Air Force seeks arise from the need to increase the effectiveness of drones. High-resolution, multispectral imaging systems and nanotechnology enhanced infrared technologies are created to increase the effectiveness of drone surveillance. In the 2009 DoD report, the Air Force provides a rationale for nano-enhanced munition: "Munitions enhanced with nanoaluminum powder are being developed to give performance and lethality enhancements in miniaturized munitions needed for the reduced volume of advanced aircraft munitions bays and for weaponization of unmanned air vehicles (UAVs)."

My insight suggests that by the 2030s, Air Force drones will account for over two-thirds of all fighter aircraft. My rationale for this is threefold.

1. Artificially intelligent (AI) machines will equal human intelligence in the period 2025–30. You may wonder if this is a wild prediction, a machine with the general intelligence of a human. It is not. Many futurists, familiar with AI, would agree,

and we will discuss the rapid pace of artificial intelligence in a later chapter.

2. With the advent of artificial intelligence that equals human intelligence and nanoelectronics miniaturization, an artificially intelligent machine will be small enough to replace a human pilot in a drone aircraft. Hence drones will be able to perform every role that manned fighter aircraft perform, including aircraft carrier launch and retrieval.

3. The U.S. citizens are weary of losing husbands and wives, sons and daughters, in remote countries for ideological reasons, regardless of the rationale. As the Wall Street Journal reported, "Americans in large numbers want the United States to reduce its role in world affairs even as a showdown with Russia over Ukraine preoccupies Washington, a Wall Street Journal/NBC News poll finds."

When we think of drone fighter aircraft, our minds often conjure images of the Predators and the larger Reapers. Both represent scaled-down versions of manned fighter aircraft. Future drones will be the size of birds and even insects, especially when the mission is for reconnaissance or assassination.

Let us now turn our attention to the U.S. nuclear arsenal, which is under the control of Strategic Command. The Air Force and Navy are the key players. Since Strategic Command is headquartered at Offutt Air Force Base in Omaha, I think it is appropriate to discuss the latest developments in missiles.

The latest development relating to the strategic missile capability is development of the intercontinental hypersonic missiles. Although the DoD is tight-lipped regarding the technology embedded in their intercontinental hypersonic missiles that approach targets at a steep, almost flat angle of attack and Mach 10 speeds (i.e., ten times the speed of sound), my view is that nanotechnology is playing a significant role. Before we discuss the technology, let us understand the mission.

Russia and China are aware of the United States' formidable

stealth technology. As a result, both Russia and China are developing stealth weapons and methods to detect an adversary's stealth weapons. To use an old cliché, it is a cat-and-mouse game. Intercontinental hypersonic missiles remove the need for stealth. Their hypersonic speed and steep angle of attack make them virtually impossible to detect or destroy with current antimissile defenses. The strategic goal of the DoD is to be able to reach and destroy any target in the world in less than an hour. The DoD claims their payload is conventional, but in light of mini-nukes, the line between conventional and nuclear weapons is becoming fuzzy. Regarding the technology, the propellant is likely using nanoparticles. My rationale is that superthermites using nanoparticles have proved to be useful in increasing the energy release of conventional explosives and propellants. Therefore, it makes sense that the DoD would consider using superthermites to boost their missiles' propulsion. Given their hypersonic speed and the need to make extremely fast trajectory corrections, there is a significant likelihood the DoD is using state-of-the-art guidance and control systems, which would suggest nanoelectronic processors. It may sound like *Star Wars*, but the United States is already successfully testing the missile. Russia and China are also attempting to develop a hypersonic missile. Russia is testing its Yu-71 hypersonic boost-glide missile and China its wu-14 rocket-boosted hypersonic glide missile. Currently, and perhaps in large part due to nanotechnology, the United States is leading in this arms race. I want to add a cautionary note. The use of nanotechnology in intercontinental hypersonic missiles is speculation on my part, but the dots appear to connect.

I think it is also appropriate to discuss mini-nukes in this section. As discussed in chapter 1, Russia, Germany, and the United States are pursuing mini-nukes, which are small nano-enhanced nuclear bombs with a minuscule amount of fissionable material. Mini-nukes can result in an explosion, depending on construction, that equates to a hundred tons of conventional explosives with almost no radioactive fallout. Based on the small amount of fissionable material and the lack of radioactive fallout, the United States may classify mini-nukes as conventional weapons.

The United States began developing mini-nukes under President George W. Bush in 2002. The exact methods to construct mini-nukes remain Top Secret. However, some nanotechnologists speculate that a high-energy nano laser could trigger a comparatively small thermonuclear fusion explosion in a mixture of tritium and deuterium. Given that the United States began research in 2002, it is likely that mini-nukes exist. Since the United States seeks to use conventional weapons in regional conflicts as well as in their intercontinental hypersonic missiles, it is conceivable that mini-nukes may fit the role, especially when seeking significant destruction.

5. Nanoweapon Thrusts of Other Nations

Other nations are developing nanoweapons. This arms race garners little in the way of daily media coverage. Yet the major players spend billions of dollars annually.

Think of this section as an Impressionistic painting. The painting may appear fuzzy because the major players are keeping their nanoweapon thrusts under a cloak of secrecy. The pigments are ground from over twenty years of media remnants, such as articles, books, and news reports. In suggestive brush strokes, the painting will surface those nations participating in the nanoweapons arms race. A few words of caution are in order. I am using the analogy of an Impressionism painting to convey that what follows is widely subject to interpretation. The Impressionism movement started because of the invention of the camera. The camera was able to reproduce pictures with excellent clarity. This meant that artists were no longer required to paint a picture to document a scene. The camera could do that with ease. This freed artists to interpret a scene, and the movement of Impressionism was born. In what follows, I provide my interpretation, much like Impressionist painters provide their interpretation. I invite you to review the salient information to form your own interpretation.

How do we determine the major players? I suggest we use two methods: follow the money and follow the patents. Let me clarify why I chose this approach.

1. You may argue that nations are unlikely to seek patents on their nanoweapons. I agree. However, I argue that a nation's capability in nanoweapons will closely align with their commercial applications, where they will seek patent protection. What I am suggesting is what I have argued all along. Nanotechnology commercial applications form a base for nanoweapons development.

2. You may wonder why I did not include nanotechnology papers in technical journals. While I agree that scientific nanotechnology publications are an indication of research, patents are a strong indicator of research output as it relates to applications. In my view, it is a stronger measure of a nation's nanotechnology prowess.

To address item 1, let us start with nanotechnology investments by nation. This is difficult to measure for several reasons. First, no common definition of nanotechnology or a nanotechnology-based product exists internationally. Second, nations seek to hide their investments in weapons. However, even with these difficulties, the U.S. government, as well as other organizations, has sought to quantify the nanotechnology investments of other nations.

Let's start with the U.S. government's estimates. According to the NNI website, nano.gov, "estimates from 2008 showed the governments of the European Union (EU) and Japan invested approximately $1.7 billion and $950 million, respectively, in nanotechnology research and development. The governments of China, Korea, and Taiwan invested approximately $430 million, $310 million, and $110 million, respectively. . . . This compares to 2008 U.S. government spending of $1.55 billion." Obviously, this is a single data point, and my research indicates it is not comprehensive. For example, there is no mention of Russia, who in 2007 successfully tested the world's most powerful nonnuclear air-delivered bomb, nicknamed the "father of all bombs," which relies on nanotechnology. This is a clear indication that NNI is intentionally omitting references to nanoweapons. However, from the 2012 NNI workshop, whose proceedings are in the public domain, we can get a more complete picture of nanotechnology global funding. One report from the workshop is from London-based consulting firm Cientifica. The report is "Nanotechnology Funding: A Global Perspective," by Tim Harper. Cientifica asserts that it compiles the data every two years, it includes government and private funding, and it is the most accurate funding picture available. After surveying what is available to the public, Cientifica's claims appear valid. These are the salient points from their report:

There is a global investment in nanotechnology approaching a quarter trillion dollars by 2015.

Governments account for $67.5 billion, and the remainder is private funding, such as corporate research.

In Purchasing Power Parity (PPP) terms, Cientifica delineates the countries making the largest investments as, in descending order, the United States, Russia, and China. Coming in a relatively distant fourth and fifth, in descending order, are the European Commission (European Union) and Japan.

This gives us a reasonably clear idea regarding the major players. From all sources, I judge the United States is clearly the largest player, but close on its heels are Russia and China.

The Cientifica report addresses another important question: How effective is the funding? To address the question, Cientifica looks at a number of factors, such as overall global competitiveness, quality of scientific institutions, capacity for innovation, levels of company spending on R&D, and "the real funding." From this, they derive the Nanotech Impact Factor, which is provided below for the top three players:

THE WOLF IN SHEEP'S CLOTHING

United States	120.41
China	98.18
Russia	98.18

This is as far as our "follow the money" approach is going to take us. Let us now address item 2, "follow the patents." These are the number of nanotechnology patents filed in the U.S. Patent and Trademark Office by the United States, China, and Russia in 2015:

United States	4,365
China	393
Russia	8

The 2015 fillings are indicative of previous years. If you look at 2015 patents filled in the European Patent Office, the top three retain their rankings with 421, 38, and 3, respectively.

Based on the above research and assuming that national spending on nanotechnology and filed patents equates directly to the development of nanoweapons, I conclude the following:

1. The United States leads in the nanoweapons arms race

2. China is a near follower

3. Russia is a distant third

Let me share my rationale for the above rankings. Clearly, based on national investment, the Nanotech Impact Factor, and the patents filed, the United States is leading. Using the similar reasoning, China and Russia place second and third, respectively.

What is surprising is that Russia appears to be a distant third. Cientifica estimates that Russia is investing more than China in nanotechnology. Why isn't Russia's investment yielding better results? One answer is "Rusnano," but that story is complex, with twists, turns, corruption, and intrigue. Let us start at the beginning.

In 2007 Russian leaders established the Russian Corporation of Nanotechnologies. This move was likely a response to emulate the effective U.S. National Nanotechnology Initiative, established in 2000. Due to both a weak Russian economy and systemic corruption, in 2011 the state corporation re-registered as an open joint-stock company, Rusnano. However, at the point of reorganization, the Russian government owned all of Rusnano's stock. Why reorganize? According to the chairman of the executive board of Rusnano, Anatoly Chubais, "Reorganization as an open joint-stock company is a transition to the greater transparency and responsibility that characterizes this organizational form. We are moving to a different plane of collaboration with the business and science communities. And that will be reflected in new high-technology projects in the Russian nanoindustry." Ultimately Rusnano seeks to sell its managing company to private investors by 2020. I speculate that the state corporation may have had issues related to corruption and interacting effectively with the business and science community, resulting in financial losses. Unfortunately Rusnano is off to a rocky start. According to Russia's Accounting Chamber, as of 2013, Rusnano had losses totaling $1.37 billion, including losses of $40 million in shell corporations (dating back to 2007), $450 million in a nonfunctioning semiconductor facility, $800 million in reserves against risky ventures, and $80 million in 2012 operational losses. Facing these embarrassing revelations, President Putin accused Chubais of poor decisions guided by CIA operatives. As a result, Chubais spoke at a news conference, admitting mistakes. "We've made more mistakes in the nanotechnology sector in Russia than anyone else. We know the pitfalls better than anyone else—so who else are you going to give your money to besides us? This is an argument that works." Chubais added, "By 2020, we've been tasked with raising 150 billion rubles ($5 billion). Believe me, it's a very ambitious task when you want to raise funds not for oil, gas, or Moscow property but for Russian high-tech, nanotechnology." However, the situation for Rusnano grew worse with time. In 2015 Putin placed Leonid Melamed, president of Effortel and of the executive board of the Supervisory Council of Rusnano, under house arrest pending an investigation related to the "waste of public money." After Melamed's arrest, five senior Rusnano managers left Russia. To sum it up, Russia's ride into nanotechnology is a train wreck and why, even after significant investment, they are running a distant third on the nanotechnology world stage.

Let's now look at how China and Russia may be applying their nanotechnology to create nanoweapons. China is the world's second largest economy with the second largest defense budget, relative to the United States. Given their economy and significant military capabilities, China presents a formidable presence on the world stage. Earlier we ranked China's nanotechnology thrusts as second only to the United States. Therefore, it should come as no surprise that, similar to the United States, China seeks to integrate its nanotechnology into its weapon systems.

On September 3, 2015, China celebrated the seventieth anniversary of the Chinese People's War of Resistance against Japanese Aggression during World War II. The Beijing military parade offered the world the first official revelation of many new weapon systems, including their latest strategic and anti-ship missiles. Missing from the parade were three of China's highly secret weapons, including:

1. The hypersonic strike missile, wu-14

This new hypersonic missile is in many ways similar to the U.S. intercontinental hypersonic missiles under development, in that it is capable of traveling up to Mach 10 (i.e., ten times the speed of sound). Its first test in early 2014 adds China to the list of countries (United States and Russia) that are testing hypersonic missiles. China launches its hypersonic missile using a booster missile, known as "boost-glide" technology, to reach the edge of space. It then reenters the atmosphere at Mach 10 to glide toward its target at a flat angle. This makes them difficult to destroy using current antimissile defense systems. The wu-14 can carry a conventional or nuclear warhead and would be effective against land and sea targets. My comments on this weapon's use of nanotechnology are similar to those mentioned regarding the U.S. intercontinental hypersonic missiles. There is a possibility that the WU-14 boost-glide technology uses nanoparticles to increase its propulsion and nanoelectronics in its guidance system.

2. Cyber warfare forces

China's cyber espionage and reconnaissance appears to be highly effective and is among their most closely guarded secrets. The United States and other countries, including Australia, Canada, and India, have accused China of cyber espionage, which China denies. China has accused the United States of cyber espionage, which the United States denies. The reality is that it is close to impossible to prove that cyber attacks are state sponsored, due to the difficulty in tracing the source of the attack. It would be reasonable to think that China would employ top-of-the-line computers to conduct its cyber warfare. This would suggest the use of nanoelectronic processors, similar to the Intel® Core™ M processor. China has a well-earned reputation regarding its ability to copy U.S. and Russian military technology. Using nanoelectronics in warfare qualifies it as a nanoweapon.

3. Ground launched antisatellite missiles

The United States relies heavily on surveillance and communication satellites to conduct military operations. China has two antisatellite missiles. According to the Washington Free Beacon:

A forthcoming report by the congressional United States–China Economic and Security Review Commission contains an entire chapter on China's military and civilian space capabilities.

The report discusses two antisatellite missiles, the SC-19 and the larger DN-2, which are meant to be fired in predetermined flight paths as a satellite passes over Chinese territory. . . . The report did not include the latest DN-3 test . . . Air Force Lt. Gen. John "Jay" Raymond, commander of the Joint Functional Component Command for Space, said at a House hearing in March that "we are quickly approaching the point where every satellite in every orbit can be threatened."

Rick Fisher, a military analyst with the International Assessment and Strategy Center, said that if the DN-3 interceptor test is confirmed, "this is the first mention of the DN-3, apparently either a modified version of the DN-2 or a new antisatellite weapon."

The complexity of targeting low and high orbit satellites suggest highly sophisticated guidance systems, potentially using nanoelectronic processors. A long held military practice is to use the most advanced technology possible, since development and deployment of a military system generally require years to decades. Using the most advanced technology during development ensures the military system will not be obsolete at the time of deployment. Therefore, it is completely reasonable that China would use nanoelectronic microprocessors. If correct, this would represent another application of nanoweapons.

In China's three new weapon systems, nanotechnology is likely playing a role. The most probable role is in nanoelectronics. Given the high commercial profile of the Intel® Core™ M processor, China is almost certainly aware of its capabilities and has found a way to obtain samples. History suggests China will find a way to copy them or obtain samples in nefarious ways to apply them to their weapon systems. When we think about the military might of China, we need to think in terms of "asymmetrical warfare." In this context, asymmetrical warfare means winning against a foe with a superior military by neutralizing the military systems that make the foe superior. For example, a technologically advanced military like the United States would be crippled if it lost its surveillance and communications satellites.

Lastly, let us discuss Russia's nanoweapons. Although we ranked Russia third with regard to nanotechnology, we should view it as a serious player. It is obvious the Russian leadership understands the importance of nanotechnology, both commercially and militarily. Russia also has a solid trade relationship with China, including China's import of Russian crude oil and weapon systems. For example, Russia sold China its s-400 Missile Defense Systems, with delivery scheduled in mid-2017. The details of the deal are murky. I conjecture that Russia is getting high-tech military weapons as part of the deal. China has hypersonic strike missile technology, successfully tested several times. Russian leadership wants that technology, and they are accelerating their development of a hypersonic intercontinental ballistic missile. I leave it to you to connect the dots.

As previously discussed, Russia is engaged in developing nanoweapons. In chapter 1, we discussed Russia's 2007 detonation of "the father of all bombs," which relies on nanotechnology to achieve its destructive power. In 2014 DefenseReview.com published an article suggesting that the Russians are also pursuing nanomaterials to improve their body armor, as well as the weapons their soldiers carry. This article makes it clear the Russians are aware of the benefits offered by nanomaterials and are pursuing nanoweapons. Although Russian nanotechnology thrusts appear behind those of the United States, during the Cold War Russians proved their ability to remain a "near follower" in military technology by using espionage and disproportionately investing a large portion of their gross domestic product in their military. In a speech to the Federal Assembly in 2013, President Vladimir Putin stated, "Russia will respond to all the challenges, both political and technological. We have all the necessary potential." Deputy Prime Minister Dmitry Rogozin, who oversees the Russian military and defense industry, made it clear that Russia will not engage in a costly arms race but will maintain military parity through "asymmetrical means." These statements are a clear indication that Russian leadership is aware that technology is critical to maintain military parity with the United States. It is obvious that they are pursuing nanomaterials to improve body armor as well as other weapons. Their timing suggests they have

THE WOLF IN SHEEP'S CLOTHING

a means to know and copy some aspects of the United States' nanoweapons thrusts.

During the mid-2000s, both Russia and the United States acknowledged building mini-nukes. Purportedly, these nuclear bombs weighed 60–70 pounds and could fit in a backpack, not a suitcase. High-ranking GRU defector Stanislav Lunev claimed Russia possessed nuclear bombs the size of suitcases, but his claims are suspect according to the FBI. With the advent of nanotechnology, there is a heightened concern that a true mini-nuke, about the size of this book with a weight around 5 pounds, is feasible. According to a 2005 MIT review article:

> Nanotechnology "could completely change the face of weaponry," according to Andy Oppenheimer, a weapons expert with analyst firm and publisher Jane's Information Group. Oppenheimer says nations including the United States, Germany, and Russia are developing "mini-nuke" devices that use nanotechnology to create much smaller nuclear detonators.
>
> Oppenheimer says the devices could fit inside a briefcase and would be powerful enough to destroy a building. Although the devices require nuclear materials, because of their small size, "they blur the line with conventional weapons," Oppenheimer says.

Does the United States, Russia, or Germany have mini-nukes? U.S. nanotechnology is highly advanced, and I speculate it is possible that the United States has developed a mini-nuke. Russian nanotechnology is primitive by comparison, and I have reservations that they have one. However, Russia has a long and deep background in nuclear technology and espionage. Therefore, while I think the probability is low, I would not rule it out completely. Germany has a robust nanotechnology infrastructure, and in my opinion, Germany has the capability to build a mini-nuke.

Let me make an important point. Both China and Russia emphasize developing an asymmetrical warfare capability. This largely stems from two difficulties both nations face:

1. Maintaining military investments on par with the United States.

2. Attempting to achieve military parity in areas where the United States has decisive leadership.

Lastly, before we conclude this chapter, let us discuss two important points:

1. A slight asymmetry in nanoweapons capability could be the deciding factor in any conflict.

2. Will it be possible to develop, deploy, and use nanoweapons in warfare without rendering humanity extinct?

In light of China and Russia's apparent strategy to pursue asymmetrical warfare, their emphasis on nanoweapons should not be surprising. Neither nation has the financial and technological resources to reach full conventional and nuclear parity with the United States within one or two decades. On the world stage, to rank as a superpower they need substantial asymmetrical military capabilities. Nanoweapons offer that potential. Even if they had only one devastating nanoweapon that the United States could not defend itself against, it could hold the United States at bay. For example, if they had weaponized nanobots capable of destroying steel and smuggled billions of them into the United States, our most formidable weapons would become a questionable deterrent. In a sense, we could be back to MAD (i.e., mutually assured destruction). Although the United States could use nuclear weapons against such an adversary, our civilization would likely crumble. Even a slight asymmetry in nanoweapons, such as the capability to kill tens to hundreds of millions of people with something as simple as toxic nanoparticles, would be a formidable deterrent. Let us also be clear about delivery of a nanoweapon. It is unlikely for an adversary to wait for the outbreak of hostilities to attempt a nanoweapon delivery. It would be easier and strategically advantageous to have the nanoweapon in place, inside the target foe's borders, prior to an outbreak of hostilities. Given the size of nanoweapons, like nanobots and nanoparticles, smuggling makes sense, and manufacturing them within the borders of a target foe makes sense. This is not a new or completely radical idea. Stanislav Lunev, the high-ranking GRU defector we discussed earlier, claims that Russia has already smuggled tactical nuclear weapons known as RA-115 "suitcase bombs" into the United States to assassinate U.S. leaders in the event of war. As mentioned earlier, the FBI questions his claims, but his assertions suggest that this type of assassination strategy was, at minimum, known in the GRU.

Let us now address the question, Will it be possible to develop, deploy, and use nanoweapons in warfare without rendering humanity extinct? Before attempting to answer it, let us understand the threat that nanoweapons pose. Nanoweapons have the capability to destabilize the balance of power, whether it is with nanoelectronics guiding hypersonic intercontinental ballistic missiles or millions of insect-sized nanobots capable of assassinating the population of a nation. If one rogue state or one "accident" provokes a nanoweapons attack that a nuclear power like the United States cannot defend itself against, then expect things to escalate quickly to the point when humanity could face extinction. Let's take two fictitious war game examples to illustrate this point.

Example 1. China sinks a U.S. aircraft carrier. A nation like China misinterprets the actions of a U.S. aircraft carrier operating in the China Sea as an attack on mainland China. China uses nanoelectronically guided WU-14 hypersonic glide missiles to attack and sink the carrier. The U.S. submarines and other warships in the carrier group receive orders to use all force necessary to neutralize the threat. Simultaneously, the United States targets all Chinese military forces, land, sea, and air that pose a threat. At this point, things escalate beyond the limits of diplomacy. Automated defensive and offensive systems engage faster than the human ability to stop the conflict. Emboldened by China's attack, North Korea unleashes a nuclear attack against South Korea and Japan. The United States uses nuclear ballistic missiles to annihilate North Korea. China's nuclear submarines, in the mid-Pacific, launch a nuclear ballistic missile attack on the United States, but the national origins of the attacks are fuzzy and appear to originate from Russian submarines, known to be operating in the area. The United States retaliates against Russia. A global conflict ensues. This is a fictitious war game example, but the military capabilities in the example exist.

Example 2. A rogue nation unleashes a nanoweapons attack on the United States by introducing lethal nanobots into every major U.S. city. The nanoweapons attack goes undetected until millions of people show symptoms and die. The United States has no effective countermeasures to halt the attack short of committing its entire military capability. The United States and its NATO allies retaliate. In a broad-spectrum attack, they use nuclear missiles against rogue states and terrorist groups deemed responsible. In doing so, the launch of nuclear weapons is "misinterpreted." Russia interprets the launch as a first strike attempt by the United States and immediately launches a nuclear retaliation. The exchange ignites a global conflict, forcing other nations like China and North Korea to engage perceived adversaries. The destruction that ensues reduces humanity to the pathetic remnants that we see in apocalyptic movies. This is a fictitious war game example. Lethal nanobots may already exist, and we will discuss them in chapter 5. Their weaponized delivery does not exist, but it is within a decade or less to realization.

The answer to the question regarding nanoweapons rendering humanity extinct is complex. We entered the nanotechnology age in 2000. Since our entry, nations have been developing nanoweapons, and humanity continues to survive. However, our current nanoweapons are primitive relative to those that will appear in the arsenals of nations in one or two decades. Will we continue to survive when those nanoweapons exist?

I left Example 2 for last. It describes a lethal nanobots attack scenario. Make no mistake. While nanobots are still in their infancy, the development pace of nanotechnology suggests we should prepare ourselves for the rise of the nanobots.

5   The Rise of the Nanobots

Something big is going on in war today, and maybe even the history of humanity itself. The United States military went into Iraq with a handful of drones in the air. We now have 5,300. We went in with zero unmanned ground systems. We now have 12,000. And the tech term "killer application" takes on new meaning in this space.

—P. W. SINGER

Nanobots are nanoscale robots. For the most part, the evolution of nanobots has its roots in warfare. Let's briefly examine their evolution.

The history of military robots dates back to the nineteenth century. In 1898 Nicola Tesla, called a mad scientist by some and the "father of electricity" by others, demonstrated his wireless remote control boat to the U.S. Navy. His pitch was that the Navy could use it to produce radio-controlled torpedoes. It must have boggled their minds, because they passed on adopting the invention. This was not the first time the United States passed on a high-tech weapon. In 1866 British engineer Robert Whitehead developed the self-propelled torpedo. The torpedo drew great interest from navies around the world, but not from the United States. However, in 1891, one event caught the attention of Americans. During the Chilean Civil War, the Chilean naval vessel *Almirante Lynch* sank the rebel Chilean armored vessel *Blanco Encalada* using a Whitehead torpedo. Clearly the torpedo had promise. That prompted the United States to commission its first torpedo boat destroyer, long after other navies had deployed almost a thousand such boats. Even in 1891, the big turnoff for the U.S. Navy was that early torpedoes were not reliable, meaning they could not stay on course to hit an adversary's ship at a distance. Although the compressed air engine of the Whitehead torpedo had a range of 1,000 yards, it was unlikely to remain on target for that distance, confining its practical use to about 400 yards. For a navy vessel to be within 400 yards of an adversary meant its almost certain demise by enemy fire. Even 1,000 yards presented perilous problems. That changed in 1896 when Austrian naval officer Ludwig Obry invented the gyroscope. This single invention became a game changer, making the torpedo stable and a long-range weapon. It would also play an important role in stabilizing ballistic missiles, but that was to come almost half a century later.

After getting a "no" from the U.S. Navy, the not easily discouraged Tesla traveled to the United Kingdom, made the same offer, and received another "no." Apparently it was a technology ahead of its time that neither government valued. However, Tesla must have made some impression on the world navies, for his thinking soon found application during World War I. Germany, for example, deployed remote control boats, *Sprengbootes*, with three hundred pounds of explosives tethered to a fifty-mile-long cable, which made the boats and operator vulnerable. Germany then adopted Tesla's wireless remote control technology to address this vulnerability. The Royal Navy deployed an aerial torpedo, consisting of a remote controlled biplane, Sopwith AT. The United States experimented with the Wickersham Land Torpedo, a tethered armored tractor meant to carry hundreds of pounds of explosives into enemy trenches, and the Kettering Bug, a bomb in the form of a small plane built to fly a predetermined route and explode at an enemy's position. However, the war ended before the United States could deploy its military robots in combat.

During the 1930s and into the early 1940s, the Soviet Union developed the first wireless remote controlled unmanned tanks, "teletanks." The Soviets deployed two teletank battalions on the eastern front during the "Winter War" of World War II, a feat not achieved by other nations until recently. The Soviet Union also deployed remote controlled boats, "telecutters," and planes, "tele-planes." Although it is not obvious these robotic weapons played a significant role during the war, the Soviet Union deserves credit for their remarkable robotic achievements.

The German Wehrmacht deployed the Goliath tracked mine, a remote controlled demolition vehicle. Its mission was to destroy tanks, disrupt infantry formations, and demolish buildings and bridges. It too represented a leap in military robotic technology, but it was not wireless. Instead, it was controlled using a 2,130-foot wire cable. In 1942 the Wehrmacht deployed the Goliath on every front, including the Allied D-day invasion of Normandy. Although the German war machine produced over 7,500 Goliaths, the weapon was largely ineffective. Its 6 mph top speed, less than five inches of ground clearance, thin armor, and vulnerable cable made it easy to disable, even by rifle fire. In fact, most of the Goliaths encountered on D-day were useless because Allied artillery fire had cut their long cables.

However, Nazi Germany did deploy robotic missiles that terri-fied London, first with their v-1 cruise missile and later with their v-2 ballistic missile. Nazi Germany launched over 8,000 v-1s at Britain, with the majority aimed at London, but less than 2,500 reached the city. The v-1's pulse jet engine resulted in a relatively slow missile, traveling about 400 miles per hour. Because they were slow with a predictable trajectory, British antiaircraft guns could destroy them en route. However, the v-2s were true ballistic mis-siles, with speeds over three times the speed of sound. They were invisible until they exploded on target. Fortunately, their deploy-ment came close to the end of the war. Nazi Germany launched 1,000 v-2s before Allied troops destroyed their launching sites.

The United States began its military use of robotic drones in 1940, with its purchase of 53 Dennymite radio-controlled aircraft from World War I British flyer Reginald Denny, a hobbyist who

2. Micro drone from DARPA. Content from DARPA Image Gallery, http://
www.darpa.mil/about-us/image-gallery.

built model airplanes and opened the Reginald Denny Hobby
Shop on Hollywood Boulevard in 1934. As World War II loomed
in 1940, threatening to engulf the United States, Denny pitched
the U.S. Army to use his remote controlled 6-hp, 12-foot wing-
span drones for antiaircraft gunnery targets. After the bombing
of Pearl Harbor and the United States' entry into World War II,
the U.S. military bought nearly 15,000 Dennymites, which they
designated the OQ-1. In 1944 the U.S. Army Air Force and Navy
embarked on Operations Aphrodite and Anvil, respectively, to
create a large bomber filled with explosives to attack highly fortified enemy positions. The weapon strategy was to let a crew fly
the bomber close to the target, bail out, and let a ground crew
wirelessly pilot the bomber. On its first mission, the bomber carried ten tons of Torpex, an explosive more powerful than TNT.
Unfortunately, the Torpex proved unstable and exploded en route,
killing the bomber's two crewmembers and forcing abandon-

THE RISE OF THE NANOBOTS

ment of the project. More successful was the Army-developed VB-I Azons, unmanned remote controlled glide bombers carrying 1,000 pounds of explosives. In 1944, 450 VB-I Azons saw combat in the Pacific and Burma.

The robotic weapons of World Wars I and II had little effect on their outcomes. However, in some cases they did cause terror, serious destruction, and loss of lives. In addition, from the standpoint of military history, they laid the foundation for military robotics.

Postwar development of remote controlled weapons slowed considerably. The newly formed U.S. Air Force felt threatened by unmanned aircraft. The Pentagon left development of such systems to the Army and Navy, which resulted in only two significant contracts. The first, in 1962, was to Ryan Aeronautical to manufacture an unmanned reconnaissance aircraft, which flew 3,435 missions over Southeast Asia between 1962 and 1975. The second, in 1979, was to Lockheed for the MQM-105 Aquila program, which the Army intended to be a small, propeller-powered reconnaissance drone. Program overruns amounting to half a billion dollars, almost twice the cost of the original program, resulted in program cancellation in 1987, with only a few prototypes produced.

Aerial drones did not play a major role during the 1991 Persian Gulf War, although the United States did use Israeli-developed Pioneer reconnaissance drones with some success. Most notable examples are the Navy's use of drones to pinpoint targets for naval bombardment and the surrender of a group of Iraqi soldiers to a drone. That event marked the first time in history that humans surrendered to a robot.

In 1995 integration of the U.S. Global Positioning System with unmanned aerial vehicles (UAVs) enabled the United States to perform reconnaissance and targeting missions with extreme precision. This was a game changer and ushered in modern drone warfare. In 1999 the United States successfully used General Atomics RQ-I Predators and Northrop Grumman RQ-4 Global Hawks in NATO air reconnaissance operations against Serbia during the Kosovo Conflict. In 2003 a U.S.-led coalition invaded Iraq to root out weapons of mass destruction. Although the United States did not find any, it has continued its military presence in the region and greatly expanded its use of robotic systems.

Since 2003 the U.S. military has been developing autonomous robotic systems, starting with the Mobile Autonomous Robot Software program. In 2005 the U.S. Army began developing the Autonomous Rotorcraft Sniper System, an unmanned autonomous helicopter carrying a remote controlled sniper rifle. Obviously the United States has an ongoing autonomous weapons research program, and it appears to be bearing fruit. In a 2014 media release, the Navy announced a technological breakthrough that will allow an "unmanned surface vehicle (USV) to not only protect Navy ships but also, for the first time, autonomously 'swarm' offensively on hostile vessels." The Navy plans to deploy the autonomous swarmboats in 2017.

The United States is not alone in developing autonomous weapons. Defense One reported, "In March 2014, the Russian Strategic Missile Forces announced it would deploy armed sentry robots that could select and destroy targets with no human in or on the loop at five missile installations." According to U.S. Deputy Secretary of Defense Robert Work, in 2015 Russia and China were heavily investing in autonomous military robotic systems. Secretary Work stated that their aim is to deploy a roboticized army, and he quoted Russian chief of general staff Valery Vasilevich as stating, "In the near future, it is possible that a complete roboticized unit will be created capable of independently conducting military operations."

In recent years, a new thrust in military robotics is emerging, namely, shrinking them. For example, the U.S. military is developing a new generation of reconnaissance drones, some the size of birds (see figure 2) and others the size of insects.

On February 2, 2016, DARPA announced that their Fast Lightweight Autonomy (FLA) program, whose goal is to develop algorithms that allow UAVs to fly autonomously, achieved a successful first flight. DARPA reports, "Sensor-loaded quadcopters that recently got tested in a cluttered hangar in Massachusetts did manage to edge their way around obstacles and achieve their target speeds of 20 meters per second." DARPA is almost certainly going to apply this technology to micro drones. Autonomous micro drones about the size of birds and insects would offer the U.S. military the ability to enter buildings, performing surveillance and potentially offensive operations. This gives a completely new meaning to "being a fly on the wall." According to DARPA's announcement, human operators would focus on "higher-level supervision of multiple formations of manned and unmanned platforms as part of a single system." The notion of multiple formations suggests a "swarm," which we will discuss in the next chapter.

If you are reading about military technology development, it is likely old news, because the military is already working on something even more cutting edge. Since DARPA is sharing their work on micro drones, I suspect it likely they are also working on nano drones and nanobots. While this may seem a leap of imagination on my part, its basis is published research. For example, on December 16, 2014, the Army Research Laboratory announced the creation of a "fly drone" weighing only a fraction of a gram. Although researchers say it may take another decade or more before the Army has a deployable fly-sized drone, I judge it is much closer, in view of advances in medical nanobots.

While the technology that enables advanced military nanobots is typically secret, advanced medical nanobots are not. Examining advances in medical nanobots will provide insight into military nanobots. Let's begin by examining two medical announcements regarding nanobots:

1. On May 15, 2015, Pfizer revealed that it is partnering with Dr. Ido Bachelet, manager of Bar-Ilan University's robot laboratory, on DNA nanobots.

2. nextbigfuturereports.com, "Bachelet has developed a method of producing innovative DNA molecules with characteristics that can be used to 'program' them to reach specific locations

in the body and carry out preprogrammed operations there in response to stimulation from the body." In this case, the preprogramming involves detecting cancer cells and delivering an existing cancer drug treatment directly to a cancerous cell, bypassing healthy cells. According to 3tags.org, Bachelet forms the nanobots from single strand DNA folded like a clamshell to hold the drug until it reaches its preprogrammed target cell. As 3tags.org's report notes, "The first DNA nanobot trial in a human subject will take place this year—in fact, it could be happening right now—on a person with late-stage leukemia. The patient is expected to die, but Bachelet believes that, based on previous animal trials, the nanobots can remove the cancer in the span of a month. If the trial goes well, we could see nanotechnology hit the public in one to five years." By "this year," they mean 2016. However, as of this writing, there have been no further announcements. This could mean several things. First, the human trial may not have taken place. Second, the human trial may not have been successful. Third, Pfizer is intentionally being quiet to enable them to continue their research without a spotlight of public and competitor scrutiny.

Bachelet and Pfizer are not the only ones using DNA nanobots. Jack Andraka, science prodigy and winner of the Smithsonian American Ingenuity Youth Award, is building nanobots out of iron oxide nanoparticles that can destroy cancer cells. Andraka makes his own nanoparticles "in a jar" and uses programmed DNA molecules to make them artificially intelligent. According to Andraka, "With the DNA you actually just buy the sequences. I tell them what sequence I want and they ship it to me and then I can just pop it on my nanorobots." He makes it sound easy, but then he is a science prodigy, and to him it probably is easy.

Other medical researchers are taking a different approach to treating cancer at the cellular level. In essence, they remove some of the patient's T-cells, which are cells produced by the patient's thymus gland. T-cells work as part of the human immune system.

After removing the T-cells, researchers alter them in the laboratory with a gene therapy to make them recognize a protein on the cancer cells. Then they inject the altered T-cells into the patient's bloodstream. There the T-cells order the cancerous cells to return to their normal configuration. If the cell has mutated too far, it orders the cell to self-destruct. Their results, reported in 2013, have been astounding, causing the cancer of fourteen out of sixteen terminally ill patients to go into remission. It is arguable whether the reprogrammed T-cells are true nanobots. I leave that to your judgment. However, I feel that the results are worth noting. It demonstrates the effectiveness of fighting diseases at the cellular level.

Nanobots are science fact. Given the current research, we could see breakthroughs in nanomedicine to treat numerous diseases, including cancer, within the next five years. If DARPA's research pans out, our next major conflict could involve micro or even nano drones. There is no doubt in my mind that all this will happen. Micro drones are close to being a reality and may already be in field tests. Nano drones will follow close on their heels within a decade. The difficulty in making autonomous micro and nano drones is cramming all the processing power into such a small package. However, nanoelectronics and DNA programming are rapidly evolving, which holds promise to make this possible.

What capabilities should we expect in military micro and nano drones? Surveillance is certain. However, the military will also want offensive capabilities. This suggests, for example, bird-sized drones that explode, taking out an adversary's command and control center. It also suggests micro and nano drones that assassinate. Imagine a micro or nano drone capable of injecting DNA nanobot "bombs," folded DNA molecules with toxins that release when the DNA nanobots sense DNA associated with genetic ancestry. It would represent the ultimate in racial or ethnic targeting.

The technology to determine DNA ancestry exists. Companies, like Ancestry.com, already offer genetic testing services. Many local pharmacies sell test kits for less than one hundred dollars. In addition, if the military had a sample of a terrorist's DNA, conceivably they could release their micro or nano drones on a "seek and destroy" mission targeted at that specific terrorist. There are DNA sequences that are more prevalent in people from one area of the world than from another area and DNA sequences that are prevalent in a specific family. With DNA targeting, it would be possible to wipe out specific people or a specific person, including their entire lineage.

Early on, I mentioned that controlling nanoweapons is a major issue. Imagine a scenario where a military unleashes "killer nanobots" with DNA targeting to kill a few specific people. Now what if their programming has a glitch? For example, suppose the military releases 10 million killer nanobots, but the DNA targeting does not work properly, and the killer nanobots kill everyone they contact. You may ask, Why would a military release 10 million killer nanobots to kill just a few people? One reason is that the location of the intended targets could be widely scattered over a large geographical area. If that military believed that their nanobots were harmless except to the few with specific DNA characteristics, it is plausible that the military would release millions of killer nanobots. Glitches have been the bane of programming from the beginning. Programming DNA is similar to programming computers. The complexity of the DNA strand makes a glitch seem entirely possible. In this scenario, with 10 million killer nanobots killing everyone they contact, the killer nanobots become a weapon of mass destruction.

In the future, the nanobots could be complex molecules in a configuration that performs a specific function. It is conceivable that the molecules could be the result of a chemical reaction, just waiting for programmed DNA strands to give them artificial intelligence. Nanoweapon factories may end up being pharmaceutical companies, manufacturing nanobots capable of curing cancer on one production line and nanobots capable of assassination on another. A nanofactory may be small enough to fit on a tabletop. This would make transporting one easy, but detecting one difficult. This suggests that nanofactories designed to build nanobots will exist not only within a nation's borders but covertly within an adversary's borders, enabling an attack from within. The superpowers of the future will require a nanoweapons industrial infrastructure capable of supplying large quantities of nanoweapons for offensive purposes. For example, if a country wants to attack the largest weapons of an adversary, they may design nanoweapons that attach themselves to steel structures and deliver a payload that causes corrosion. For an attack like this, billions to trillions of nanobots will be required. In effect, nanobots will attack a nation's largest weapons, like aircraft carriers, by borrowing a tactic from nature called "swarming."

6 The "Swarm"

The fiercest serpent may be overcome by a swarm of ants.

—ADM. ISOROKU YAMAMOTO

The smallest of weapons, nanoweapons, are able to attack and destroy other nations' largest weapons by borrowing a useful tactic from Mother Nature: swarming. A single ant or bee is typically innocuous. However, consider army ants and killer bees. Their predatory swarming, typically termed "attacks," causes sizable animals, including humans, to avoid them.

We see evidence of swarming's effectiveness in almost all lifeforms. Ants, bees, and wasps all exhibit swarm behavior to attack larger prey. Termites swarm to attack buildings by eating the wooden beams that support them. Dragonflies, beetles, butterflies, and moths migrate in swarms. Locusts swarm to devour crops, at times covering 100 square miles and consuming over 100,000 tons of plant matter in a single day. About 1,800 bird species swarm (i.e., flock) to migrate long distances. Some fish swarm (i.e., school) to attack, such as piranhas. Tuna, herring, and anchovies school to defend against predators and to enhance their probability of finding food and mates. Even humans swarm. Early humans formed tribes and hunted in groups to find and kill larger prey. In the modern era, humans form civilizations and work together for mutual benefit. Swarming is without a doubt one of nature's most useful tactics. It enables smaller animals to compete with larger animals.

If you have any doubt about the effectiveness of swarms to harm their intended targets, you need look no further than the 1918 Spanish flu pandemic, involving the H1N1 influenza virus. Viruses, among the smallest of all life-forms, infect by swarming to destroy the host's healthy cells. The 1918 pandemic infected 500 million people and left 100 million dead, which was about 5 percent of the world's population at the time. Viruses are not nanobots in the traditional sense. Yet they are similar in size to nanoparticles. For example, the H1N1 is about 80–120 nm in diameter and usually spherical, and they attack at the cellular level. Nanobots can use swarming tactics to disable larger objects. To accomplish this, though, they must work together with an increased level of intelligence, similar to the way army ants work to achieve their goal. Let's illustrate this with a simple example.

Imagine nanobots capable of seeking and destroying weapons-grade radioactive material. One such nanobot would do little damage, perhaps destroying a few atoms of the nuclear material. Billions to trillions could render the nuclear arsenal of a nation useless. If the nanobots were artificially intelligent to mimic swarming bees and their target was weapons-grade radioactive material, releasing them in close proximity to nuclear missile silos might be sufficient. Bees, for example, sense pollen and nectar-laden flowers by their bright colors, the patterns on the petals, their aromatic scents, and their electric field. The bee's ability to sense a flower's electric field is a recent finding. Apparently this capability allows a bee to distinguish between fields of different flowers. It even enables the bee to determine whether other bees have recently visited a flower. To locate fields of flowers, bees send out scouts. If a scout scores, it signals the hive. Some scouts do this by marking the route with aromatic flower oils, others by leading their hive mates. Honeybees communicate to the hive by "dancing." The dancing is a form of intricate communication. The nanobot could use a similar strategy. Once a nanobot scout finds the nuclear material, it could emit a signal. A nanobot's size would suggest that the signal would be extremely weak, perhaps just reaching the next nearest nanobot. Their tactic to communicate to the swarm might be similar to the "fourth general order" of the U.S. military. The U.S. armed forces, regardless of branch, maintain "The Eleven General Orders." The fourth general order is "to repeat all calls from posts more distant from the guard house than my own." By employing nanobot to nanobot signaling, in a matter of seconds, the entire swarm may know the location of the nuclear material. To destroy it, they may simply dislodge the material, atom by atom, until it is a pile of radioactive dust. This would render it useless as a weapon. Imagine trillions of nanobots in a swarm and countless swarms. In a short while, perhaps measured in hours, all land-based nuclear missiles would be useless.

While under attack, the humans who control the missiles may know only that there is a malfunction at their location. They would report it to their superior officer, who would forward the message up the chain of command. By the time the message reaches the top command echelons, there may be no recourse except to alert the "boomers," the nuclear missile–carrying submarines. Even if the boomers are operational, the perpetrator may not be obvious. Which nation do they target?

You may wonder, How can nanobots get into the nuclear missile's payload? Surely each missile is in a silo, covered by steel blast doors, to provide some hardening against nuclear weapons. The radioactive payload is inside an airtight container. Unfortunately, the exact details of nuclear weapon storage are classified. However, here is a scenario that addresses the question. The deposition of thin films requires a vacuum. There is an old adage that dates back to Aristotle, "Nature abhors a vacuum." Aristotle argued that material surrounding a vacuum would inevitably find a way to fill the void. From my experience as a thin film process engineer, I can confirm that Aristotle was correct. Creating and maintaining a vacuum to deposit thin films is difficult. The vacuum pumps continually fight miniscule leaks. I would venture to say that the meaning of "airtight" is relative. Every man-made vacuum has a miniscule leak, perhaps only noticeable using a helium leak detector. Since the helium molecule is smaller than the oxygen molecule, helium is able to traverse the miniscule passages and reveal the leakage points. Conceivably, nanotechnologists may develop nanobots with a diameter smaller than even an oxygen molecule and able to take advantage of any miniscule passage. Perhaps such nanobots would have a rod shape, similar to the DNA molecule, which is only 2 nanometers in diameter and about 3 nanometers long. Like virus, they may be airborne, carried by winds, or travel like snakes on the ground. Once a scout finds the radioactive fuel, it will lead the swarm to its location. If the minuscule passage is in the nanoscale range, the scout will enter and the remainder of the swarm will follow.

To destroy large structures requires incredibly large numbers of nanobots. I cited the DNA molecule as an example of a complex structure that only has a 2-nanometer diameter and a 3-nanometer length. If Mother Nature can make such structures, it is conceivable that eventually nanotechnologists will, too.

In the last chapter, we discussed several medical breakthroughs to fight cancer. In each case, it involved a "swarm" of medical nanobots to treat the patient. The technological nanobots in two of those examples used modified DNA molecules to provide the nanobots with artificial intelligence. In this, we are seeing a blend of nanotechnology and biology. It would not be surprising to see this methodology extended to weaponized nanobots, perhaps to the point they could distinguish organic tissue from inorganic elements. To target a specific inorganic substance is simpler than analyzing chemical composition. The weaponized nanobots seek only a preprogrammed match. In the case of radioactive material, it may use the ionizing properties of radiation to detect it, similar to how a Geiger counter works. Perhaps, at the nanoscale, the collision of radioactive particles with members of the nanobots swarm may be the indicator. Nanobots attacking radioactive material present a unique challenge. The radioactive material is emitting particles and gamma rays, which could strike the nanobots like a bullet and destroy them. To nanobots, collision with emitted particles would be similar to nuclear missiles hit by high-energy projectiles. This also argues for enormous numbers of nanobots for such a mission. They would need to overcome their heavy losses to successfully carry out the mission.

Although destroying a nation's land-based nuclear weapons would be a devastating blow, nations like the United States, Russia, and China have nuclear missile submarines, boomers. Therefore, an attack on only the land-based missiles would not neutralize the threat of retaliation. Even if nanobots succeeded in assassinating the bulk of people and leadership of a nation, the nuclear retaliation would reduce the suspected adversaries to radioactive rubble. Clearly there would be no winners. For an attack to succeed, an adversary would have to disable a nation's ability to retaliate, which would be extremely difficult. Let's understand why this would be so difficult.

At the height of the Cold War, the United States employed the nuclear triad, which included land-based nuclear missiles, nuclear missile submarines, and airborne bombers carrying nuclear bombs and cruise missiles. Today the United States, Russia, and China still employ land-based missiles and nuclear submarines. This makes attacking them, even with nanoweapons, a no-win scenario. In a sense, the doctrine of mutually assured destruction still applies. However, understand that terrorists and rogue states may still launch such an attack, literally hoping to slip under the radar or even cause a war between nuclear powers. Unlike a nuclear missile attack, the perpetrator of a nanoweapons attack is difficult to detect. This makes the world an even more dangerous place than it was during the nuclear age.

Does any government possess swarming nanobots capable of destroying a large military target? The answer is a qualified yes. The Faculty of Life Sciences and the Institute of Nanotechnology

& Advanced Materials at Bar-Ilan University in Israel published "Universal Computing by DNA Origami Robots in a Living Animal." According to the abstract:

> Here, we show that DNA origami can be used to fabricate nanoscale robots that are capable of dynamically interacting with each other in a living animal. The interactions generate logical outputs, which are relayed to switch molecular payloads on or off. As a proof of principle, we use the system to create architectures that emulate various logic gates.

In effect, they created robots capable of interacting with each other and performing functions, including sensing. Although their paper discusses the nanoscale robots' behavior inside a living animal, they performed proof of concept outside a living animal. This is getting close to our scenario. The capability of "interacting with each other" suggests that if properly programmed, they could swarm and communicate. Since the interactions generate "logical outputs, which are relayed to switch molecular payloads on or off," it is conceivable that the payload could be a material that attacks a specific structure. It is interesting to note that their paper uses two terms to identify its content, "DNA computing" and "DNA machines." This suggests that the authors understand the potentially wide applications their work may enable.

The reason I qualified my answer is that the Israeli scientists are directing their work toward nanomedicine, with the hope of curing diseases like cancer, not nanoweapons. However, there are many historical examples of making commercial and medical products into weapons. The most significant challenges to turning the nanobots into weapons in this example are twofold:

1. Making them capable of performing in the harsh environments found outside living animals, including wind, rain, dirt, and pollutants, to name just a few.

2. Making the payload perform the desired function, such as disassembling atoms of a larger structure.

These are significant obstacles, but we do know that Mother Nature is able to do it routinely with viruses. Some infectious flu viruses can live up to eight hours on a surface, such as a book or doorknob. It is conceivable nanotechnologists will eventually learn to mimic nature and develop nanobots with similar or even more robust qualities.

I have little doubt that some nations like the United States and China are working on the problem. However, their closeness to a solution is a well-kept secret. It appears, in this regard, that nano-medicine is ahead of nanoweapons. To be effective weapons, nano-bots would need to work under the typical conditions found in real life, be controllable, and be intelligent enough to perform their mission. Mother Nature has given us proof of concept regarding real-life survival of nanoscale organisms outside living animals. We know it is possible, and eventually nanotechnologists will learn how to do it. Addressing the control issue may mean, on a rudimentary level, just limiting the geographic location and fuel that powers the nanobots. Providing just enough fuel to operate within a geographic location may be all the control necessary for the first generation of such nanoweapons. When they run out of fuel, they cease to function. Unlike living creatures, they need not be self-replicating and capable of refueling. The big challenge is intelligence. DNA is unlikely to provide the complete answer to nanobots working outside an animal's body. If the attack is on a man-made structure, its structures do not have DNA or easily recognized cells. They are made of atoms and molecules. Therefore, weaponized nanobots will require a greater level of artificial intelligence to detect and destroy man-made structures. Human intelligence allows educated adults to differentiate between a missile and a baseball bat without chemically analyzing them, even though they have roughly the same shape. If we had that level of intelligence in nanobots, it would simplify targeting. Will we reach that level of artificial intelligence?

Part 2

THE GAME CHANGERS

7 The "Smart" Nanoweapons

> There are lots of examples of routine, middle-skilled jobs that involve relatively structured tasks, and those are the jobs that are being eliminated the fastest. Those kinds of jobs are easier for our friends in the artificial intelligence community to design robots to handle them. They could be software robots; they could be physical robots.
>
> —ERIK BRYNJOLFSSON

Most people use a "smart" phone to make a telephone call or even watch a movie. Today's smart phones make *Star Trek's* communicators seem primitive by comparison. My iPhone has numerous applications. With the touch of a screen icon, I can find out the weather or baseball scores, send an instant message, and perform a number of other functions. What makes the iPhone smart? Artificial intelligence. The word "smart," in this context, is synonymous with the phrase "artificial intelligence." If you own a smart phone, it is artificially intelligent and has more computing power than NASA had when it landed two men on the moon.

Have you noticed military announcements and commercial advertising rarely use the phrase "artificial intelligence"? Instead, the military states they use "smart bombs." Commercial advertisements state you can own a smart anything, from a phone to a microwave oven. Consumers routinely buy cars that park themselves, microwave ovens that have one button for popcorn, and thermostats that regulate room temperature and humidity relative to the outside weather conditions. The reason why we do not often hear the phrase "artificial intelligence" is that generally we do not focus on technology. We focus on function. For example, we might say we have a self-parking car. We do not say, "We have an artificially intelligent car." Few would understand our meaning. When artificial intelligence (AI) enables a product's function, we describe the product by its function, not its technology. You may have read an article about Google's self-driving cars. Notice, we are describing the product by its function, not its technology, which is artificial intelligence. Now that we are discussing all these applications, you might wonder, How does artificial intelligence work?

The technological wonders, like smart phones and other smart products, use a computer, often only a microprocessor chip, running specific programs to provide instructions for a product to perform its functions. Each function requires a different program, and each manufacturer uses a different approach to accomplish the same function. Similarly, AI researchers can use different approaches to accomplish the same task. In fact, AI is a multidiscipline field, and the approach to make a product artificially intelligent largely depends on the background of the researchers. The AI that enables a product to perform a specific goal is termed an "intelligent agent." In fact, many textbooks will define the field of AI as the study and design of intelligent agents. For example, if you play chess on your phone, you are playing against an intelligent agent, a software program that runs on the microprocessor that makes expert-level chess moves in response to your moves.

At the heart of all AI applications are a computer and a software program. The combination of the computer's processing power and the software's capabilities determines the machine's intelligence. No computer currently has the equivalent processing power (i.e., the speed at which a computer can perform an operation) of a human brain, but the Chinese Tianhe-2 supercomputer comes extremely close. It is difficult to define how close because there is no widely accepted definition regarding the processing capability of the brain. However, even if Tianhe-2 equaled its processing power, it still would not equate to human intelligence. It lacks the software to enable it to think as humans think. If it could think like a human, we would say it has "general artificial intelligence," as opposed to being an intelligent agent.

You might wonder, How will we determine when a machine has general artificial intelligence equal to a human? In 1950, mathematician and computer scientist Alan Turing devised just such a test. Many people are now aware of Alan Turing from the movie *Imitation Game* as the man who broke the Nazi Enigma code. The Turing test is widely accepted among AI researchers to represent the minimum test that a machine would have to pass to equal human intelligence. It works like this. Imagine a person in the same room as a computer. Imagine that the person will type questions and comments to engage the computer in a conversation. The computer will answer and comment. If a third person is unable to tell which portion of the conversation is the computer versus the person, researchers argue the computer passes the Turing test and is generally artificially intelligent. One point to keep in mind is that the answers to the questions do not have to be correct. Just as humans do not know the answers to all questions, the computer does not need to know the answers to all questions. The computer's reply only needs to appear human.

To date, no computer/program has successfully passed the Turing test, but some have come close. The most notable is the Eugene Goostman software program, developed by a team of programmers from Princeton, Saint Petersburg, and Kiev in 2001. In 2014, in a contest held by the Royal Society and marking the sixtieth anniversary of Turing's death, thirty judges conversed independently with the Goostman program and a human. Ten judges thought Goostman was human. I do not think the test is conclusive, due to the constraints. The information given to the judges led them to believe that Goostman was a Ukrainian thirteen-year-old boy. Thus, even if some of the answers appeared wrong or simplistic, the judges would be tempted to attribute that to the subject's age and nationality. A lot of controversy surrounds the interpretation of the test.

When will artificial intelligence equal human intelligence? Most AI researchers predict this will occur in the time frame 2025–30. How was this projection determined? The way we arrive at this is by assessing the current capability of AI technology and then using "Moore's law" to determine where it will be in the future. But what is Moore's law? In 1975 Intel founder Gordon Moore observed that "the number of transistors incorporated in a chip will approximately double every twenty-four months." This means that the new chip would have twice the density of circuits and cost about the same as the old chip. A colleague of Moore made another observation. If Moore was correct, the processing power of a computer would double every eighteen months. This observation took into account that smaller circuit features meant faster processing. Using Moore's law, futurists are able to predict when a combination of computer and software will equal human intelligence. However, this is not the whole story. Most researchers believe that Moore's law only applies to computer hardware, not software. Although there are different opinions, in my observation software capability appears to increase linearly. In addition, I view Moore's law as broader than just applying to integrated circuits. In my view, Moore's law is an observation regarding humanity's creativity. It can be used to project improvements in technology when that technology field is highly funded, like integrated circuit technology. The reason that it historically did not apply to computer programs is funding. Most of the funding went to developing the computer. A smaller fraction went to developing the software to run the computer, but that is beginning to change. Given that we are on the cusp of a computer almost passing the Turing test, with Goostman, I judge one decade should be sufficient to realize that goal. Just consider this. In ten years, if integrated circuit technology continues to follow Moore's law, the processing power of a typical computer will be over 100 times greater than today's computer. I think that generation of computers will pass the Turing test.

The U.S. military uses artificial intelligence in its weapon systems. In fact, they have been funding AI from the beginning, and it has been giving them a return on their investment. The most notable are smart bombs, which garnered an enormous amount of television coverage during the Iraq wars. However, when computer intelligence equals human intelligence, it is reasonable to believe that weapons, like drones, will become autonomous. Humans may specify mission parameters, but the computers will carry out the mission. This is similar to how a commanding officer gives the personnel in his command the mission goal, and they carry out the mission. In the case of drones, even the use of missiles may be autonomous. Warfare as we know it will no longer exist. Instead of humans performing life-threatening missions, autonomous AI robots, which we will term smart robots, will perform those missions.

There is a lot of negative press coverage regarding fully autonomous weapons. We will discuss it by first defining terms. Fully autonomous weapons relate to weapon systems that can select and attack targets without human intervention. The concern is that the computer will carry out the mission without injecting judgment, which every human pilot must exercise. When pilots make mistakes, they are held accountable. This accountability requires pilots and their mission controllers to exercise extreme care before using a weapon. All political and military ramifications require careful judgment. For example, during the Cold War it was typical for the Russians and the United States to breach one another's air space to measure the quickness of their respective responses. Typically several jets would be scrambled to intercept the intruder, who would detect them on radar and return to international air space before their arrival. Each side was aware of the other's intention and acted accordingly. Without judgment, such intrusions could result in an exchange of missiles.

Does a computer with intelligence equivalent to a human have judgment equivalent to a human? There are numerous articles on the subject. Some argue yes and some no. I have read many of them, and I conclude that no one knows. The question is hypothetical, since we do not have computers equivalent to human intelligence. However, the reality is that Russia is deploying a fully autonomous weapon. For example, according to Defense One, "In March 2014, the Russian Strategic Missile Forces announced it would deploy armed sentry robots that could select and destroy targets with no human in or on the loop at five missile installations." China is likely to follow. To maintain military parity, the United States is also likely deploying such weapons. The autonomous weapons currently deployed do not have artificial intelligence equivalent to human intelligence. Therefore, we can conclude that they lack human judgment. The real issue is whether without human judgment these autonomous weapons will kill innocent people or even ignite a war. Unfortunately, only the future will provide those answers.

There are also numerous ethical considerations surrounding autonomous weapons. Here is a short list:

1. Will autonomous weapons act in accordance with the international humanitarian law?

2. Will autonomous weapons act proportionally in response to an attack?

3. Who will be accountable for the actions of autonomous weapons?

The largest issue regarding questions 1 and 2 is preventing harm to noncombatants. In general, the U.S. military complies with international humanitarian law. However, can the United States, or any nations, ensure compliance when deploying autonomous weapons? Question 3 is not only an ethical question but also a military question. If an autonomous weapon malfunctions and commits an act of war, does this require retaliation? For example, consider this scenario. Russia has an autonomous missile defense system surrounding Moscow. Suppose it malfunctions and destroys a U.S. aircraft in international airspace. Should the United States retaliate? If the destroyed aircraft is part of an autonomous drone formation, it is conceivable that the formation would attack the Russian autonomous defense system. What happens next may be the start of World War III.

The use of autonomous weapons is in its infancy. There are many military and moral questions regarding their use. However, nations are developing and deploying them even as I write these words. To my eye, it appears they make the world more danger-ous because these first-generation autonomous weapons lack human judgment. Will later generations have the capability of human judgment? At this point no one can definitively answer the question. That is why the United Nations is working with its members to ban the use of autonomous weapons, but that pro-cess is just beginning, and it appears to be almost at a standstill.

Let us use this information to understand what this may mean regarding nanoweapons. In my view, we can formulate three crit-ical implications regarding nanoweapons in the 2030s:

1. Computers, with AI equal to human intelligence, will design nanoweapons, especially electronics and robots. Humans will define the nanoweapon parameters, and the computer will per-form the design with low human and computer interactions.

2. Nanoweapons will become autonomous, including weapon systems like drones and nanobots.

3. A nation's military prowess will equate primarily to its nanoweapons capabilities and less on its nuclear weapons capa-bilities. Thus nanoweapons will define the balance of power.

To understand the basis of these implications, let's discuss them in detail. The first two points are already happening. Nanoweap-ons and nanoelectronics are designed using computers. This is not new. The more sophisticated a weapon system or integrated circuit, the more we rely on computers to assist with the design. As computers become more capable, this trend will increase. The term that describes this methodology is "computer-aided design," or CAD. We are also seeing a trend toward more autonomy in weapon systems, like the Navy swarmboats we discussed in chapter 5. The reason for this trend is that computers can function faster than humans and typically with greater expertise. The rationale for the third point is multifold. Let us consider a nanobots attack versus a nuclear weapons attack.

A nuclear attack aimed at destroying a nation requires the use of missiles. Military powers are capable of detecting a nuclear missile launch. Indeed, with the use of satellites, the United States can detect when a potential adversary begins fueling their land-based nuclear missiles. In essence, it would be difficult for any nation to launch nuclear missiles and avoid detection. This would invite a counterstrike. In the case of a nuclear attack on the United States, all NATO nations would consider it an attack on them as well. The retaliation would be devastating. In contrast, a nanoweapons attack that involves the release of billions of autonomous AI nanobots would be hard to detect prior to the attack. Their AI capability may allow them to attack only when all nanobots are strategically in place. In a matter of hours, a nation may have no leadership or even a clue as to the nature of the attack. The stealth and lethal capabilities of a nanoweapons attack suggest that nations with smart nanoweapons will overshadow other nations with less advanced nanoweapons and even nations with nuclear weapons. The same applies to weapon systems that rely on nanoelectronics, such as hypersonic ballistic missiles with nuclear warheads. A hypersonic missile attack could occur faster than a nation could respond or determine the adversary. Nuclear submarines, the "boomers," would not know which nations to target. Even if they do launch their nuclear missiles at likely perpetrators, the adversary would be expecting retaliation and would be ready with antimissile defenses. The boomers themselves would give away their positions by launching their missiles and become targets for "killer subs."

It is conceivable that a nation could develop autonomous smart nanobots capable of traversing space and water, like viruses, to attack a nation's land-based, air-based, and sea-based weapons.

THE "SMART" NANOWEAPONS

Again, they could lie dormant until all nanobots are in place before performing their assigned military tasks. For example, a military commander could assign nanobots to find and destroy enemy submarines, missile silos, and bombers with nuclear weapons. This would render the doctrine of mutual assured destruction useless. I recognize that today this example sounds like science fiction. However, given sufficient funding, nanotechnologists could develop such nanobots. No law of science prohibits their existence. The task would be monumental, similar in scope to putting a man on the moon, which was once also science fiction. In addition, given that such nanobots may require years to carry out their assigned mission, it would likely require nanobots to generate their own power. For example, they could use solar power or wind power to charge a nanoscale battery. Similar to nuclear-powered submarines that are able to generate their own power for decades, self-powered nanobots would pose a viable threat indefinitely.

Even if nations control their nanoweapons, terrorists and mad scientists may still initiate an attack. Consider this scenario. A disgruntled nanoweapons scientist decides to seek revenge for some perceived unjust treatment. The scientist makes billions of autonomous smart nanobots at home and takes them to a reservoir in a suitcase. There, the scientist releases the nanobots, which contaminate the drinking water of a major metropolis. Within a few weeks, people drinking the contaminated water begin to show signs of illness. However, at this point, it is too late for countermeasures. Millions of innocent people begin to die. The nation is in shock, and its citizens live in fear. The government wants to act, but it is not clear what they should do. Even if they discover the mad scientist, beyond prosecuting him, what can they do to prevent a similar attack?

One nanoweapons attack could ignite a global conflict. Consider the mad scientist scenario, but replace him with a terrorist group. If terrorists launch autonomous smart nanobots on a nation, that nation may retaliate against any nation known to harbor such terrorists. Such a widespread retaliation may ignite a global conflict as nations attempt to defend themselves. In the fog of war, it may be impossible to determine who is doing what to whom. A nation may face a "use it or lose it" situation. This means that if they do not use their weapons, they may lose them. In such a scenario, any outcome is possible.

I have examined many scenarios related to autonomous smart nanoweapons attacks. Every scenario led me to the same two conclusions:

1. Nanoweapons are inherently dangerous, even to the nations that deploy them. One incident, intentional or accidental, could ignite a global conflict.

2. Nations deploying nanoweapons will garner military respect, but they will also garner high scrutiny, even to the point of paranoia. We may find ourselves in a new and even more dangerous cold war.

By 2045, most researchers and futurists in artificial intelligence predict that computers will exceed the combined cognitive intelligence of humanity. Google's director of engineering, Ray Kurzweil, calls this point in time the "singularity." In my opinion, singularity level computers will design nanobots capable of self-replication, and they will do this with minimal human assistance. Once self-replicating nanobots become a reality, just a few nanobots gaining entry to a submarine would eventually result in the swarms necessary to destroy it. I have found that the easiest way to think about nanobot attacks is to liken them to biological agents. In effect, self-replicating nanobots become an artificial life-form, and we become their gods.

8 The Genie Is Loose

> I would like to say that the only thing that would preserve security would be to lock everybody up, and when they decided to leave to shoot them and be done with it. That is the only way you could have perfect security.
>
> —GENERAL L. R. GROVES

Almost all technologically advanced countries are pursuing nanotechnologies, and some are pursuing nanoweapons. Indeed, there is a nanoweapons arms race among the United States, Russia, and China. This new arms race has garnered almost no publicity, and like nanoweapons themselves, it is invisible to most of humanity. We could appropriately term it "the invisible arms race." However, make no mistake. This new arms race is real, and each nation engaged is investing billions of dollars to have nanoweapons that qualify it as a superpower.

As of this writing, the United States appears to be winning this race. However, if history has taught us anything, it is that maintaining a military edge is fleeting. For example, the United States was the first country to have atomic bombs, which it used on the Japanese cities of Hiroshima and Nagasaki in 1945. By 1949, the Soviet Union had detonated their first atom bomb. The Soviets were able to replicate the United States' most destructive weapon via espionage and their own engineering. Most Americans were shocked by how quickly the technology secrets of the atomic bomb were in the hands of another nation. However, the U.S. Energy ırtment, which published a thirty-six-volume *Manhattan Dis-History*, reveals that even during the development of the bomb re were 1,500 leaks of classified information:

> Since September 1943, investigations were conducted of more than 1,500 "loose talk" or leakage of information cases and corrective action was taken in more than 1,200 violations of procedures for handling classified material. . . . Complete security of information could be achieved only by following all leaks to their source.

The Manhattan Project, as it is commonly termed, was Top Secret. This raises a question, Why were there so many leaks? When most people think about the development of the atom bomb, it typically conjures images of a handful of scientists working in secret. The truth is completely different. There were 130,000 people working on the atom bomb. It was a monumental project. In 1945 dollars, it cost nearly $2 billion. That equates to about $26 billion in 2016 dollars. There is an old saying, attributed to Benjamin Franklin, "Three can keep a secret, if two of them are dead." The reality is that with 130,000 people involved and spies everywhere, keeping the Manhattan Project a complete secret was impossible. The United States managed to stay just ahead of other nations. Even General Groves, the Manhattan Project leader, understood this: "Now, I think that really when it comes to military security and the relative advantages of giving out information and not giving it out, what we are interested in from the military standpoint is the relative movement, you might say, of ourselves and other nations. It isn't so much how fast we progress; it is the relative motion of the two."

This is the situation we are in today. The United States has invested over $20 billion as of 2015, managed through the NNI, and we have nearly a million workers in an estimated trillion dollar nanotechnology industry. According to a 2016 report by the director of national intelligence, James R. Clapper, "China continues to have success in cyber espionage against the United States government, our allies, and United States companies." In this report, delivered to the Senate Armed Forces Committee, Clapper said, "Russia and China continue to have the most sophisticated cyber programs." Given this insight, isn't it likely that both Russia and China can maintain a near parity position regarding the United States' nanoweapons? It is also likely the United States does likewise via the Central Intelligence Agency. Military secrets are hard, if not impossible, to keep for long durations. I fully expect that any nanoweapons deployed by any nation will soon be deployed by other nations. History bears witness to the validity of this assertion.

According to the Stockholm International Peace Research Institute, as of 2014, nine nations have nuclear weapons: the United States, Russia, United Kingdom, France, China, India, Pakistan, Israel, and North Korea. Factually, though, any industrial nation could develop a nuclear weapon within a few years if it sought to do so. The how-to information can be found on the Internet. Nations with the technological expertise of Germany and Japan could not only develop nuclear weapons but within a decade also have arsenals on parity to the United States and Russia. Treaties, like the Nuclear Non-Proliferation Treaty, entered into force on March 5, 1970, have not stopped non-signatory nations like Pakistan, Israel, and North Korea from developing nuclear weapons. In general, treaties have only been marginally effective. If we learn from history, nanoweapons will also proliferate.

Let's take a look at which nations are likely to have offensive nanoweapons. My list of "Nanoweapons Offensive Capability of Nations" (NOCON) has three categories:

1. Nanoweapons Nations—Nations that already deploy nanoweapons and have robust ongoing development programs

2. Near Follower Nanoweapons Nations—Nations that are developing or deploying nanoweapons but are one or more generations behind other nations

3. Nanoweapons Capable Nations—Nations capable of developing nanoweapons, but who have chosen not to do so

The question now becomes, How do we determine which nation to place in each category? We will use the time-proven method of "following the money." However, we will also examine three additional factors:

1. A nation's ability to commercialize nanotechnology

2. A nation's focus on military prowess, as judged by their military spending

3. National alliances that enable nanotechnology and nanoweapons sharing

Appendix 2 explains a systematic process and rationale for assigning a nation to a specific category. It requires correlation of a nation's nanotechnology capabilities, military budget, and military alliances. However, let me admit at the outset: although I have invested countless hours in constructing the NOCON list, it may be wrong. There is no standard way to construct the list. For those interested, I encourage you to read appendix 2 to gain deeper insight and formulate your own conclusions.

Nanoweapons Offensive Capability of Nations

This is how I view the Nanoweapons Offensive Capability of Nations:

1. Nanoweapons Nations

United States is the clear leader in developing and deploying offensive nanoweapons. Close alliances with NATO nations like the United Kingdom and France may prove synergistic in developing nanoweapons.

China is a near follower that understands the importance of nanoweapons, deploys them as they become available, and is ramping up their nanoweapons programs as rapidly as their economy allows. Alliances with Russia and North Korea may prove synergistic in developing nanoweapons.

Russia comes in as a distant third that has gotten a relatively poor return on its investment in nanotechnology, mainly due to corruption and poor management. Its alliance with China may prove synergistic in developing nanoweapons.

The United Kingdom is the United States' closest ally. In fact, military strategists characterize their relationship as special. The level of cooperation between them with regard to military planning, execution of military operations, nuclear weapons technology, and intelligence sharing is unparalleled among major powers. There is every reason to believe that this special relationship includes nanoweapons sharing. In addition, the United Kingdom is a nuclear power and a NATO nation. It would be in the best interest of the United States to ensure that the United Kingdom is a fully capable military ally, including its ability to deploy nanoweapons. However, this is not a one-sided relationship. The United Kingdom would be able to share any advances in nanoweapons it makes resulting from its membership in the European Framework Programmes for Research and Technological Development. From an economic viewpoint, including trade and commerce, and from a military viewpoint, the United Kingdom is the undeclared fifty-first state of the United States. I do not make this statement lightly. The United Kingdom and the United States have been close allies for well over a hundred years, and their special relationship has been forged in numerous military and political conflicts: World War I, World War II, the Korean War, the Cold War, the Falklands War, the Gulf Wars, and now the war on terror

2. Near Follower Nanoweapons Nations

France is probably developing nanoweapons via their Centre National de la Recherché Scientifique and their participation in the nanotechnology development thrusts of European Framework Programmes for Research and Technological

Development. They are a NATO nation and a nuclear power. French nuclear weapons, like those of the United States, United Kingdom, and Russia, are on high alert, ready to use on short notice. Their nanoweapons would likely result from their commercial applications, for example, using nanoparticles for automotive catalytic converters, and their ever increasingly close alliance with the United States. Some military analysts suggest that France is replacing the United Kingdom as the United States' closest ally. I do not agree, but I do think that France equals the United Kingdom in military capability. It also politically aligns with the United States regarding defeating Islamist militants and opposing Russian aggression. In fact, in 2013 French president François Hollande launched a military intervention in Mali to defeat Islamist militants.

Germany is likely developing nanoweapons. As of 2015, there are about 1,000 German companies engaged in the development, application, and distribution of nanotechnology products. Those companies account for about 70,000 jobs. In general, Germany's nanotechnologies thrusts are increasing. Historically Germany is a signatory of the Treaty of Brussels, which forbids them to possess nuclear, biological, or chemical weapons. Germany is also a signatory of the Non-Proliferation Treaty and agrees not to develop or deploy strategic nuclear weapons. However, as a NATO nation, Germany, along with Belgium, Italy, the Netherlands, and Turkey, receives tactical nuclear weapons from the United States, which also controls their use. Given Germany's strong nanotechnology infrastructure and constraints regarding nuclear, biological, and chemical weapons, it is logical that they would seek to have a strong position in nanoweapons. I judge it is likely, for example, that Germany is developing specific nanoweapons, such as mini-nukes. Germany is also a member of the European Union and participates in the nanotechnology development thrusts of European Framework Programmes for Research and Technological Development.

South Korea is the world's fifteenth largest economy and has nearly 30,000 U.S. troops stationed within its borders. South Korea is an ally of the United States and an important trade partner. In addition, South Korea's military budget is almost as large as the entire GDP of North Korea, its hostile neighbor. It is likely that:

- U.S. troops in South Korea have the latest generation of conventional weapons and tactical nuclear weapons to thwart a nuclear strike by North Korea.

- South Korea is developing offensive nanoweapons, especially related to the threat that North Korea poses.

3. Nanoweapons Capable Nations

Japan focuses its nanotechnology on commercial, industrial, and medical applications. Based on the 1960 Treaty of Mutual Cooperation and Security between the United States and Japan, Japan is a protectorate of the United States and an important trading partner. Strategically, it makes sense for Japan to focus on its continued economic growth and on strengthening its relationship with the United States. Most military strategists suggest that Japan would like to upgrade its relationship with the United States from protectorate to ally.

India primarily focuses its nanotechnology on commercial, industrial, and medical applications. India is a nuclear power. It is actively pursuing new ballistic missile, cruise missile, and sea-based nuclear delivery systems. Given their low military budget and modest nanotechnology capabilities, it is doubtful they have a robust nanoweapons program.

Saudi Arabia did not show up in Cientifica's report. Based on that, it would be reasonable to believe it is not pursuing nanoweapons. However, the 2016 edition of *The Military Balance*, published by IISS, estimates Saudi Arabia's 2015 military budget at $81.8 billion, making it the third largest military budget in the world behind only the United States and China. It is also a close ally of Pakistan, sharing commercial, cultural, religious, and political interests. Pakistan is a nuclear power. Currently Saudi Arabia's focus is on remaining one of the world's largest oil exporters. Their foreign policy appears guided by maintaining relations with other oil-producing and major oil-consuming countries. However, the Middle East is a hotbed for religious conflict and wars related to the region's oil resources and access to the Persian Gulf. For example, one major reason Iraq invaded the small nation of Kuwait in the first Gulf War was to gain direct access to the Persian Gulf. In a practical sense, Saudi Arabia has a large defense budget and aligns itself with Pakistan to deal with their perceived threat from Iran. However, that is not the whole story. The Middle East is the most militarized region in the world. Saudi Arabia's relationship with the United States has drawn criticism from radical Islamic groups throughout the region. Given Saudi Arabia's military budget, its precarious position in the Middle East, and its alliance with Pakistan, it would be reasonable to believe it would acquire nanoweapons as they became available. Today, most arms sales are to Middle Eastern nations. I am not suggesting that Saudi Arabia is developing nanoweapons. I am suggesting that their oil wealth and defense budget would enable them to buy some of the best nanotechnologists in the world to develop offensive nanoweapons or even buy offensive nanoweapons that become available in the military market, if they perceive it in their national interest.

The desire to acquire offensive nanoweapons will, I judge, strongly guide international relations starting in the late 2020s. With the advent of artificial intelligence that rivals human intelligence, smart nanoweapons will become weapons of choice. There are seven reasons for this:

1. Nanoweapons in the late 2020s will be less expensive than nuclear weapons.

2. They will be easier to deliver than nuclear weapons.

3. Detecting nanoweapons manufacturing will be difficult.

4. The destructive power of nanoweapons will rival nuclear weapons and offer the potential to be more selective, minimizing collateral damage.

5. The use of nanoweapons will appear more ecofriendly compared to nuclear weapons.

6. Detecting a nanoweapons attack will be difficult until significant damage occurs.

7. Determining the source of a nanoweapons attack will be difficult.

Items 1–7 essentially guarantee the proliferation of nanoweapons, and even small nations may wield significant military might. The definition of a superpower will change. It will focus not solely on a nation's nuclear arsenal and delivery systems but also on its nanoweapons arsenal and delivery systems. To my eye, this guarantees a new cold war.

In particular, I expect nations in the Middle East to be among the first to buy their way into becoming nanoweapons nations. I also expect the emergence of cold wars in the late 2020s and early 2030s:

1. The East versus West: China and Russia will square off against the United States and other NATO nations, like the United Kingdom and France.

2. The Middle East: Nations within the Middle East, and militant groups within those nations, will square off against each other for the same historical reasons that have engulfed that part of the world in conflict for thousands of years.

3. Korea: North Korea will square off against South Korea, as it has for over half a century.

4. Terrorism versus the rest of the world: Radicalized Islamist militant groups will square off against the United States, the United Kingdom, France, and Russia, which they see as having exploited the Middle East to control the flow of oil.

Going forward, I will refer to the nanoweapons cold wars as the new cold war, the sum of all hostilities, propaganda, and threats between nations and terrorists, short of open warfare.

Item 4 represents a new kind of cold war. The adversaries are not a nation within a specific geographical border. They are Islamic terrorists, widely spread throughout the Middle East and other parts of the world, who interpret tenets of the Quran and the Hadith to justify their acts of terrorism. But that is not the whole story. Two studies of Islamic terrorists, one in the United Kingdom and one in France, found almost no connection between religious piety and terrorism. What is it that drives them? There are numerous answers, and opinions differ. Here are two profiles that I find provide insight:

1. Olivier Roy, a professor at the European University Institute in Florence, Italy, profiles global terrorists as:

a. Coming to avenge what is going on in their country.

b. Lacking religion before being "born again" in a foreign country.

c. Coming from "de-territorialized backgrounds. . . . For instance, they may be born in a country, then educated in another country, then go to fight in a third country, and take refuge in a fourth country." (Roy also judges this lack of identity with nationality as responsible for the success that terrorist groups have in recruiting members throughout the world.)

d. Believing that jihad (a war against nonbelievers) is permanent and not linked with a specific territory.

2. Afghan pathologist Yusef Yadgari's 2007 study of 110 Afghanistan suicide bombers discloses that 80 percent had some kind of physical or mental disability.

Numerous authorities and volumes profile terrorists. These two, to my eye, capture the essence of most of the information. I conclude that the cold war with terrorists presents the greatest danger. Even if the suicide bombers represent only a tiny fraction of all terrorists, they are a serious threat that could turn a cold war into a full-out conflict.

As of this writing, numerous nations have thwarted terrorists' plots to gain nuclear weapons. However, in the late 2020s and early 2030s, nanoweapons will be available on the black market. Numerous manufacturers will make them. Unlike nuclear weapons, if some go missing, they will be harder to find than nuclear weapons. They will not give off a telltale radiation signature. I am concerned that it may be possible for suicide terrorists to gain access to offensive nanoweapons via the black market. Another issue is the portability of nanoweapons. A disgruntled employee in a nanoweapons factory may be able to find a way to steal them. One briefcase filled with nanoparticles may be enough to wipe out a metropolis. I believe authorities will apprehend the perpetrator, but only after the fact. By that point, millions of people may be dead and many others dying. Whether the disgruntled employee is a lone wolf or a mad scientist will not matter. Given the large workforce of the nanoweapons military industrial complex, the probability of this occurring is high.

9 Fighting Fire with Fire

Be stirring as the time; be fire with fire;
Threaten the threatener and outface the brow
Of bragging horror.

—SHAKESPEARE, *King John*, 1595

Defensive Nanoweapons Scenario—The White House knew something was wrong because U.S. satellites indicated that Russia was fueling ICBMs at the Plesetsk missile base located in Mirny, about five hundred miles north of Moscow. President Arnold James was not overly concerned. There were no other reports of Russian offensive activity. He knew that Russia historically launched missiles from Plesetsk to test their missile technology. However, protocol required him to contact the Russian leadership via the Moscow-Washington hotline, a secure computer link established in 2008 over which the two nations exchanged email messages. Text messages were preferred because speech could be misinterpreted. Leaders sent their messages in their native language. Each side translated the messages at the receiving end. The first official use of the original hotline occurred in 1963, following the assassination of President Kennedy. At this time the hotline was a teletype machine. In 1991 the Kremlin and the White House also established a direct phone link. Having learned from the Cuban Missile Crisis, which required six hours or more to exchange "official" communications, each nation wanted the ability to connect quickly, especially to avoid misunderstandings that could ignite a nuclear war.

The president ordered the first email sent at 8:43 a.m. EST. It read, "Dear President Kozlov, we have detected missiles being fueled at Plesetsk. Please advise intentions. Respectfully, Arnold James, President of the United States. April 23, 2035, 8:43 a.m. EST." The message was intentionally short and each word diplomatically chosen.

Any use of the hotline was automatically classified "Eyes Only—The President" and deemed urgent. Normally the United States would receive a reply within minutes, even if the reply merely acknowledged receiving the message. At 9:00 a.m., after receiving no reply, President James decided to call the Kremlin directly. Before making the call, he had Secretary of State John Callahan and Secretary of Defense Carol Baker join him and his chief of staff, Michele Powers. The phone rang six times. That was strange, as each nation had people ready to answer the phone 24/7.

Finally, a strained voice answered, "This is Stepanov." Prime Minister Stepanov spoke fluent English.

"Hello, Mr. Prime Minister, this is Arnold James." The phone was on speaker, but the president did not mention it.

"Yes, Mr. President."

"Is everything all right?"

"No, Mr. President. Everything is not all right."

"We have detected that you are fueling your missiles at Plesetsk."

"Yes, Mr. President. We are taking defensive measures."

The atmosphere in the Oval Office became tense, and everyone moved slightly closer to the speaker on the president's desk.

James kept his tone calm and even. "What defensive measures? Against whom?"

"We do not know. We are still gathering intelligence."

"Is President Kozlov available?"

"No, Mr. President. He is dead."

This took the president completely by surprise. He lowered his voice. "My condolences. . . . When did he die?"

"About one hour ago, along with four members of Parliament."

"We deeply regret your losses. Do you know what is causing them?"

"Yes, we are under attack. Nanobots are responsible. I am speaking to you from the Kremlin Defense Ministry Room." This was a highly fortified three-story structure, the equivalent to the U.S. presidential bunker. Stepanov continued, "We have to take defensive measures."

"I understand. I assure you that we are not responsible."

"Yes, Mr. President." Stepanov's tone was condescending, and the phone went silent for ten seconds. To President James, the seconds seems like minutes, and Stepanov's condescending tone concerned him. He hit the high alert button on his other desk phone and then said, "I give you my word we are not responsible and will assist in any way we can. I will call you back shortly." The call ended.

The high alert button automated DEFCON 2 via executive order, meaning that nuclear war was imminent. Eight members of the Secret Service immediately entered the Oval Office and without exchanging words grabbed the president and others. With guns drawn, the agents took them to the White House bunker. This high drama was protocol, but extremely unusual. Within twenty-five seconds, they were locked inside the bunker.

Based on Stepanov's tone, the president was not sure who the Russians held responsible. He had no alternative but to take defensive measures. He contacted General Arthur, the head of the Joint Chiefs. He and his staff had moved to the Pentagon bunker.

"General, what the hell is going on?"

"We don't exactly know, Mr. President."

"Then tell me what you do know."

"At 8:29 a.m. our satellites detected that the Russians were fueling their ICBMs at Plesetsk. Two of our killer subs also reported that the Russian boomers [*slang for ballistic missile submarines*] they were following are sinking. Other Russian boomers are acting erratically, almost as if they are not under control. None are at missile launch depth. Our forces remain fully operational."

"Thank you, General. Keep the line open and listen in. I am going to call Stepanov back."

"Yes, sir."

The president picked up the hotline in the bunker, and Stepanov answered almost immediately. The president wanted to avoid any appearance of hostility and decided to be completely open with him.

"Mr. Prime Minister, we have some information for you."

"Yes?"

"Our surveillance indicates your missile subs are acting erratically. Two appear to be sinking." The president paused to give Stepanov time to process the information. "Are you aware of this situation?"

"We are."

James persisted. "What are your intentions?"

"We will defend ourselves."

"Mr. Prime Minister, you are now the president of the Russian Federation. I urge you to not launch your missiles."

"We must defend ourselves."

"I understand, but let us help."

"Help? How?"

"We have countermeasures that will neutralize the threat Russia is facing."

"What countermeasures?"

"Defensive nanobots that neutralize offensive nanobots."

"We know about your defensive nanobots, but what good is that going to do us now?"

Stepanov's acknowledgment surprised President James. Defensive nanobots were classified above Top Secret as a special access program. *How did the Russians learn about our defensive nanobots?* There was no time for speculation. "We can't undo what some adversary has done to you, but we can stop further losses. Together we will find the parties responsible and bring them to justice."

"What do you propose?"

"We need to fire hypersonic missiles on Moscow, Plesetsk, and every major Russian city and military installation. Those missiles will carry our defensive nanobots."

"You want to fire missiles on Russia?" Stepanov's tone was disbelief.

"Yes. With warheads containing defensive nanobots."

"How can we be sure this is not a trick to launch an even greater attack?"

"You must trust me. I have put our military on DEFCON 2 to ensure we don't suffer a similar fate." Frustrated, the president said, "Look, if we were responsible, wouldn't we just launch our hypersonic missiles and continue the attack? We know that Russia has no defense against our hypersonic missiles and can't detect them until they reach their target. Why would we be talking right now if our intent was to destroy Russia?"

"Yes. What you say makes sense." There was about a twenty-second pause as Stepanov weighed his options. "We must defend ourselves to show we are still a superpower. We intend to fire our missiles at ISIS concentrations in the Middle East."

"I think that would be risky. If China or we misinterpret their trajectories, we will have to respond."

"If Russia is to be no more, then ISIS is to be no more."

"Please let me confer with my people for about fifteen seconds."

"Yes, Mr. President."

President James hit the mute button and asked, "General Arthur . . . your thoughts?"

"We can destroy Plesetsk or even their missiles in flight. Some Russian subs, bombers, or mobile launch pads may still launch a devastating counterattack on the continental United States. I cannot guarantee that we can fully destroy all possible incoming missiles."

"I understand. Please hold, General."

"President Stepanov, I urge you to let us help you." The president paused, hoping for Stepanov to reply, but after about six seconds he asked, "Have you let China know about your intentions?"

"No."

"We need to do that." The president signaled to Secretary Callahan to get China up to speed.

Callahan quietly got up and accessed another phone to have his office put him through to China's ambassador to the United States. Within a minute Callahan was discussing the situation with Ambassador Chong. In tandem, President James continued his conversation with Stepanov.

He reiterated, "We don't want any misunderstandings. We have Ambassador Chong on the line."

"Greetings, President Stepanov," said Chong in a respectful tone. "I am contacting my leadership. I urge you not to provoke China."

"We must all do what we must do." Stepanov seemed resigned to accept the hand that fate had dealt him.

"I have a proposal that may resolve this issue," said President James.

"Yes, Mr. President."

"Fire one missile that we can track and let us fire our hypersonic missiles. The one missile will demonstrate your military might."

The look on the people's faces in the bunker was a mixture of horror and disbelief. The president of the United States had just suggested that Russia fire an ICBM with a nuclear warhead. Chong was silent.

"How are you going to help?" Stepanov was confused.

"Let us neutralize the nanobots attacking you."

Stepanov believed the attack was a terrorist attack, retaliation for Russia's combat presence in the Middle East. He also thought that the United States and China would not risk nuclear war with Russia via a nanobots sneak attack. Stepanov was a graduate of the prestigious General Staff College of the Russian Federation's armed forces and a highly respected moderate in the Russian leadership. His feelings toward the United States and China were friendly. He carefully weighed several scenarios and reached a conclusion. "Okay. We will do as you suggest. You can fire your hypersonic missiles on Russia." After pausing again, Stepanov continued with emotion choking his voice. "Let us hope for Russia and the world that your plan works." The call ended.

"Mr. Chong, please alert your leadership not to overreact to this nuclear demonstration," James said in a matter-of-fact tone. James knew that China's leaders were likely already on the line and listening. Not waiting for a reply, President James continued, "I will need to put you on hold." President James hit the mute button.

"Mr. President," General Arthur spoke up, "our satellites detect Russia has launched a single ICBM. Its current trajectory suggests Syria is the target."

"Where in Syria?" James was concerned that coalition troops, mostly the United States, United Kingdom, and Russia, could be causalities of the nuclear detonation.

"Can't specify exactly . . . hold on . . . looks like it will detonate in the Syrian desert in the Al-Hamad."

"Do we have any coalition troops in that region?"

"No. It's desolate. There's nothing there, not even ISIS."

James knew that the Russian president did not want to risk a nuclear war. By targeting the Al-Hamad desert, there would be virtually no casualties. However, like firing a warning shot, it would send a message.

He took Chong off mute and said, "I'm sure that your people are tracking Russia's nuclear demonstration."

"We are tracking Russia's missile. We see it is targeting Syria."

"Yes, Mr. Ambassador," said Callahan in a conversational tone. "We think it will detonate in the Al-Hamad. No Chinese interests will be threatened."

"We will watch the situation carefully." Chong's voice was stern.

"We will watch with you," replied Callahan and put Chong on mute again. Russia's close proximity to Syria cut the missile's time to target to about seven minutes. Tension was evident on every face in the bunker. Finally, the missile's warhead detonated in the Al-Hamad.

General Arthur said, "It's a small tactical nuke, about 10 kilotons . . . smaller than the Hiroshima bomb. Fall-out should be minimal."

James took Chong off mute and said, "It was a small nuclear detonation to send a message to ISIS."

"Yes, we confirm that, but we will continue to watch this situation carefully." Chong paused, probably getting orders from his leadership, and then continued, "China will hold the United States responsible for any attack by Russia on China's assets."

"Please assure your leadership that we are doing all we can to contain this situation," replied James. He knew that China was only saber rattling, as was their custom.

"I hope we understand each other," replied Chong, and the call ended.

The president ordered Secretary Baker to fire the hypersonic missiles carrying the defensive nanobots. The missiles were always in a state of readiness, and targeting took only minutes. Baker told General Arthur to carry out the president's orders.

President James spoke into the speakerphone. "General, how long before our missiles reach Moscow?"

"First missiles from our subs will arrive in seventeen minutes. Moscow is priority 1. Our launches from South Dakota will take fifty-four minutes."

"How long for them to take out the killer nanobots?"

"Hold on, Mr. President. We're making that calculation."

About a minute passed, and General Arthur's voice came through the speakerphone. "We estimate complete eradication will require four hours after release of our defensive nanobots."

"What about the people infected?"

"The defensive nanobots will penetrate their skin and perform a 'seek and destroy' mission. Most should begin to feel better in less than an hour, but they must expose themselves to our defensive nanobots."

"What about the Russian subs?"

"If they have been infected, it may be too late for them."

"Pipe our satellite coverage of Moscow to my bunker. I want to observe."

One of the bunker's monitors came to life and displayed a satellite image of Moscow. They kept the image centered on the Kremlin with a Magnification that represented about 1,000 feet.

"Okay, General, stay on line. I'm calling Stepanov."

Stepanov answered immediately.

"Yes, Mr. President?"

"Our first missiles will be over Moscow in about fifteen minutes. They will detonate over the city. It will appear more like fireworks, but warn your people of falling missile debris."

"We have already sounded the sirens. Everyone is taking cover."

"It's important to make sure everyone gets exposed to our nanobots. That means you will have to flush your bunker with air from outside."

"What will happen to us?"

"Most people will feel better in less than an hour. However, our calculations indicate it will take four hours to destroy all the killer nanobots once our defensive nanobots arrive."

"What about our submarines?"

"If you can reach them, order them to surface and open their hatches. We will target missiles to deliver defensive nanobots to their locations."

"And if we cannot reach them?" Stepanov's voice was both sad and frustrated.

"I will ask my people." James hit the mute button, putting Stepanov on hold, and said, "General Arthur, what do you suggest?"

"Hold on, Mr. President, and give us a few minutes to confer."

"Okay. I am going to talk to Stepanov while you confer. Signal my phone when you have an answer."

James took Stepanov off mute. "President Stepanov?"

"Yes, Mr. President."

"We are working on the problem. Have you reached any of your subs?"

"Yes, five so far. They have an order to surface and open their hatch in four hours."

"Good, we'll locate them and . . ." The sight of hypersonic missiles detonating over Moscow interrupted James's train of thought.

Stepanov spoke. "We have detected your missiles detonating over Moscow. We will open our venting to the outside as soon as your defensive nanobots start to work."

"Good." James knew that Stepanov wanted to see if the defensive nanobots were effective before he completely surrendered control. The light on James's phone signaled that General Arthur was ready with an answer. He addressed Stepanov. "Can I call you back in about fifteen minutes?"

"Of course." Stepanov voice seemed to signal relief. James ended the call.

"Yes, General, what do you have for me?"

"I'm afraid it's bad news followed by worse news."

"Go ahead, General."

"The bad news is that any subs not under control will eventually sink or collide with something. Even if they surface, they will become a navigation hazard."

"And the worse news?"

"Sinking subs may contaminate the area with radiation, or a nuclear warhead may detonate."

"Okay, General, now tell me what we're going to do about it."

"Our best option is to sink the subs ourselves."

"How about the radiation?"

"We'll only sink them by using nanobots that target the con tower. Then we'll salvage the warheads."

"Can you guarantee that?"

"No, Mr. President, I cannot. I can only say that in the limited time we've had to work the situation, this is the best scenario we came up with."

"Okay, General, stay on line. I'm calling Stepanov."

James placed the call, and Stepanov immediately answered.

"President Stepanov . . ."

"Yes, Mr. President . . ."

"There is no easy answer to our problem. My people say that our best chance is to sink your out-of-control subs and salvage the warheads."

"Yes, we have come to the same conclusion." Stepanov paused and took a breath. "But any warheads recovered are the property of the Russian Federation. They must be returned to us."

"Whatever warheads we recover will be returned to Russia. You have my word."

"Thank you, Mr. President."

"I also want to thank you for detonating your tactical nuclear warhead in the Al-Hamad."

"You are welcome. Hold, please. I am getting reports that members of our Parliament are recovering. It seems your defensive nanobots are working."

"Good. Moscow should be completely safe against further attacks."

"How long do your defensive nanobots last?"

"A long time." President James did not want to let them know that the nanobots were self-sustaining and would last indefinitely, unless turned off remotely by the United States.

"I understand." Stepanov knew by James's answer that the defensive nanobots were going to last indefinitely, but did not feel this was a good time to discuss that point.

"My people are telling me that your subs are surfacing."

"Yes, we amended our orders."

"We'll take it from here."

"Mr. President?" Stepanov's voice filled with emotion. "You are a friend to Russia. We will not forget what you have done for us."

"President Stepanov, I hope this marks a new and closer alliance between our great nations . . . and we will find out who is behind this."

"Thank you, Mr. President. We must continue our dialogue. There is still much to do."

Although the preceding scenario is fictitious, it is entirely plausible.

If history has taught us anything, it is these four truisms:

1. It is in humanity's nature to engage in war. Modern humans evolved about 200,000 years ago. However, evidence of our ancient ancestors using spears dates back to 400,000 BC.

2. Every war gives rise to weapons with increasingly greater capability for destruction. Many experts estimate that the current nuclear arsenal is sufficient to render humanity extinct.

3. For every offensive weapon deployed, an adversary soon develops a defensive weapon to nullify it. For example, the deployment of tanks during World War I in 1917 spurred the invention of a German antitank weapon in 1918. It was a large rifle firing a 13mm solid bullet that could penetrate the tank's armor and destroy the engine or kill the occupants. Every offensive weapon, right through modern times, appears to have a defensive counterpart. Even ICBMs are destroyable in flight by antiballistic missile systems, like the Russian A-135 and the U.S. Ground-Based Midcourse Defense system.

4. Military secrets are extremely difficult to keep.

If you accept these truisms, I suggest that the history of nanoweapons will be no different.

Similar to our fictional scenario, the most effective countermeasure against nanoweapons, would be defensive nanoweapons, such as smart defensive nanobots. The size of smart offensive nanobots and their dispersal make them impossible to attack using conventional or nuclear weapons without significant collateral damage. For example, it is not possible to use nuclear weapons to stop smart offensive nanobots from attacking the leadership of a nation without killing the leadership and all people in proximity. If the smart offensive nanobots covered a ten-mile radius, that might mean killing millions of people and doing irreparable harm to the environment. Again, thinking about smart offensive nanobots as a virus is a good model. How do you eradicate the virus without eradicating the affected population?

We need to stop offensive smart nanobots at their own level using defensive smart nanobots. In this context, by "smart nanobots" we mean they are autonomous and have AI functionality. Unfortunately, using defensive smart nanobots to fight offensive smart nanobots becomes problematic. Here are some of the problems:

• The defensive smart nanobots may not be completely effective, just as antiviral medications are not effective on all viruses. Even today, we don't have a cure for the common cold.

• The defensive smart nanobots may prove harmful to some people, just as an antiviral medication can have serious side effects for some people.

• The offensive smart nanobots that are able to evade the defensive smart nanobots become, in a sense, resistant, much the same way that influenza viruses can become resistant to antiviral medications.

• If the defensive smart nanobots have an extensive longevity, their AI code could eventually become corrupt, causing them to become a danger to animals and humans.

The major issue with using smart nanobots as an offensive or defensive weapon boils down to control. The initial nanobots may become uncontrollable following their release. One approach to address the control issue is to limit their power source so that, like normal life-forms, they cease to function after some period of deployment. Another approach would be to have them respond to a homing signal, similar to the way some bacteria, known as magnetotactic bacteria, follow the Earth's magnetic field to migrate in a specific direction. I judge that nanotechnologists will build in methods to control nanobots, but the real test will be the modern battlefield, which could end up being your or my city. In addition, any controls must be sufficiently sophisticated to ensure that an adversary does not hack them. If our four truisms remain in effect, it is likely any type of secret external control signal will eventually be in the hands of an adversary. This is part of the problem.

Let us imagine we are in the mid-2030s, and some nations have smart nanoweapons. Each nation will seek to make its offensive nanoweapons resistant to another nation's defensive nanoweapons. Each nation will seek to make control of their nanoweapons so sophisticated that other nations will not be able to compromise them. The new nanoweapons arms race, like all arms races before it, will become a cat-and-mouse game. As one class of nanoweapons becomes vulnerable, another class will be developed that eliminates the vulnerability. In a sense we will be back to the Cold War arms race between the former Soviet Union and the United States, except that now there are numerous adversaries, some known and some unknown. Some will be familiar. Some will be new. Most concerning will be the unknown and nontraditional nanoweapons superpowers. Today a nuclear superpower needs a significant amount of weapons-grade uranium, a huge chemical processing facility to yield highly concentrated plutonium (Pu-239), a complex mechanism to make it into a bomb, and a missile delivery system. These are detectable by satellite surveillance and their radiation signature. A nanoweapons superpower may only require a facility about the size of a house, and the delivery system might be a suitcase. Traditional military prowess may not count for much.

If our artificial intelligence and nanoweapons technology continues on its current course, by the mid-2030s a number of nations will be deploying offensive and defensive nanoweapons. The defensive nanoweapons will give rise to a new generation of more resistant offensive nanoweapons. It is a vicious cycle. As Virgil put it in Latin, *aegrescit medendo*, which translates as "the cure is worse than the disease." Unfortunately, if any nation or terrorist group achieves an asymmetrical nanoweapons advantage where they can launch an attack and simultaneously neutralize any counterattack, the world will find itself at a tipping point.

Part 3

THE TIPPING POINT

10 The Nanoweapons Superpowers

The superpowers often behave like two heavily armed blind men feeling their way around a room, each believing himself in mortal peril from the other, whom he assumes to have perfect vision.

—HENRY A. KISSINGER

The year is 2050. The world has experienced two "technological singularities." In this context, we will define a technological singularity as a point in time when technology profoundly alters human evolution. Here are the technological singularities that will occur by 2050:

1. Artificially intelligent machines will exist that exceed the combined cognitive intelligence of humanity. These machines, which we will term "Singularity Computers," will be in the hands of governments and possibly a few wealthy companies, like Google. Singularity Computers will increase the rate of technology development in every field, including nanoweapons, and will initially appear to interact well with their human handlers. They will work on any projects submitted to them, as well as numerous projects they deem require address. Singularity Computers will first emerge in the early 2040s, and their superintelligence will yield results that benefit humanity. Here are some examples:

• Pharmaceuticals to address serious human illnesses, like antivirals that cure AIDS, the common cold, and influenza.

• Gene therapies that extend human life and reverse the effects of aging.

• Self-connecting stem cells that repair spinal and brain injuries.

• Smart prosthetics to replace lost limbs that link directly with the central nervous system and are stronger and more agile than the recipient's original limb.

• Implantable image sensors that link directly with the central nervous system and enable the blind to see with greater acuity than humans with normal vision.

• Artificial organs, printed on 3-D printers, using organic "inks" that replicate the recipient's tissue down to the specific DNA.

• Nanoelectronic processors that make commercial laptops into supercomputers. Methods to upload the consciousness of deceased humans, where the person appears to be alive "inside the machine," happily living in virtual reality.

• Brain implants that augment intelligence, capable of wirelessly interfacing with the Singularity Computers to retrieve any stored knowledge.

• Self-replicating smart nanobots (SSN) that can be programmed for various functions and manufactured using a combination of existing pharmaceutical and nanoelectronic companies. This gives rise to the second technological singularity.

2. The SSN (self-replicating smart nanobot) will have the potential to render humanity extinct. These SSN have completely changed every aspect of human existence. They mine raw materials and build factories to manufacture products to satisfy nearly all human needs. They provide medical treatments for previously untreatable conditions, like inoperable brain tumors. They offer unparalleled military security, able to nullify threats in the planning stages. Singularity Computers, which continue to remain under human control, interface with SSN wirelessly and

direct their activities, including self-replication. Life in technologically advanced countries, with Singularity Computers and ssn, is like the biblical Garden of Eden. Humans want for nothing.

In this new reality, Singularity Computers will eventually perform all research and development, as well as run robotic factories to cost-effectively fulfill humanity's needs, even our food. Problems that have plagued humanity since its emergence as a species will appear solvable. The focus of humans in countries with Singularity Computers will be leisure.

Because of the technological singularities, new superpowers will emerge characterized by the prowess of their Singularity Computers and nanoweapons, especially their ssn capabilities. It will become apparent that superpowers enjoy an extraordinary quality of life. It also will be apparent that Singularity Computers are able to develop extremely effective nanoweapons, especially ssn. Nuclear weapons will still play a role, but it will be an entirely different role than they played historically.

These technological singularities will become extremely disruptive on the world stage in three ways:

Disruption 1: New, and in some cases unknown, superpowers will emerge. It will be unclear regarding which nations are the new superpowers, with both Singularity Computers and effective nanoweapons, and how they rank relative to each other. The stealth aspect will obscure their identities. However, a reasonable judgment regarding the new superpowers and their 2050 rankings is possible by using my list of "Nanoweapons Offensive Capability of Nations" (NOCON) delineated in chapter 8, the military alliances likely to exist in 2050, and the following assumptions:

• "Moore's law," in its most general form, continues to apply to technology evolution.

• Computer programming, starting in the early 2030s, begins to follow Moore's law as computers with self-learning capabilities become the norm.

• Humanity continues to pursue its current course in artificial intelligence and nanoweapons technology, free from regulation.

Before we proceed, let me provide a few words of caution. While I feel the projections are correct, I acknowledge they rest on the validity of the NOCON, projected military alliances, and the specific assumptions delineated. I leave it to you to formulate your own judgment using the information provided. There are a number of ways to envision the world in the year 2050. I think you will find this methodology interesting and logical, but I accept that other methodologies may yield an even more accurate vision.

Here are projections of how the nations will rank, in descending order, regarding the new superpower status:

• United States will be the first nanoweapons superpower. Its prominence as a superpower is directly proportional to its economy, which enables it to outspend its closest rivals, like China and Russia, by a factor of four to ten, respectively. However, China and Russia maintain near parity via espionage and their own nanoweapons development. The United States will be the first nanoweapons superpower to deploy self-replicating smart nanobots in all aspects of human endeavor, including commercial, industrial, medical, and military. The SSN represent a new prosperity engine, unparalleled in any previous industrial revolution. SSN also represent weapons far more devastating and flexible than nuclear weapons. People of the world both admire and fear the United States as the most visible and potent superpower.

• China will be a nanoweapons superpower and will not attempt to hide its capabilities. Its economy will be second only to the United States'. In tandem with its growing economy and military might, China will become "westernized," meaning that it will adopt the economic and political systems of Europe and North America. Initially China's economy flourished based on being the world's cheap labor source. However, by the 2050s robotics will change the dynamics of manufacturing and remove the need for human labor. In China's case, their first real experience with capitalism and a quasi-democratic government occurred when Hong Kong reverted to their control in 1997. It became obvious that the system worked, and China's leaders will extend the Hong Kong model to mainland China. China's deeply felt need for a strong military will continue. It will seek military parity with the United States, which is the world's de facto gold standard of economic wealth and military might. China will continue to emphasize computer technology and eventually evolve Tianhe-2 to a Singularity Computer, Tianhe-3. The stealth aspects of nanoweapons will keep much of China's capability hidden from the world, with the exception of the United States, which will use its SSN capability to monitor every nation, even its closest ally, the United Kingdom.

• United Kingdom will be a nanoweapons superpower. Most nations will suspect that the UK's relationship with the United States enables this status, but its complete capability will remain hidden. The UK will become a nanoweapons superpower out of necessity. The United States will need a nanoweapons superpower ally they can trust. The special relationship between the United States and United Kingdom will enable joint military planning, execution of military operations, nuclear weapons technology sharing, computer technology sharing, nanoweapons sharing, and intelligence sharing. In addition, the United Kingdom will play an even more important role as a NATO nation than it currently does. For reasons that we will discuss shortly, nations will be highly reluctant to allow Singularity Computers to interface with each other. However, given the special relationship between the United States and the UK, I judge that they will allow their Singularity Computers to interface, ensuring that each nation remains united to the other.

• France will become a nanoweapons power but not a superpower. France will continue to seek a closer alliance with the United States and politically align with it against rogue states and terrorist groups as well as Russian and Chinese military aggression. France will become a nanoweapons power by:

> Increasing their alliances with the United States, which I expect will include limited computer technology and nanoweapons sharing.

> Focusing on developing nanoweapons via their Centre National de la Recherché Scientifique and participating in the nanotechnology development thrusts of European Framework Programmes for Research and Technological Development.

Given sufficiently high cold war tensions, France may expand its role in NATO to include committing its nuclear and nanoweapons. French nuclear and nanoweapons, like those of the United States, United Kingdom, Russia, and China, will be on high alert, ready to use on short notice.

• Germany will become a nanoweapons power but not a superpower. Blocked initially by the Treaty of Brussels and later by the Non-Proliferation Treaty, Germany was unable to develop or possess strategic nuclear weapons. Therefore, their military prowess was limited and their defense highly dependent on the United States, which under the U.S.–NATO nuclear weapons sharing provisions provided Germany with tactical nuclear weapons under U.S. control. However, World Wars I and II provide ample evidence of Germany's need to be a world-stage military power. Based on their robust nanotechnology infrastructure and their strong commercial applications, Germany will develop effective computer and nanoweapons technology and hold a position similar to the United Kingdom's today. It will not have complete parity with the United States, but it will have sufficient nuclear and nanoweapons to be a strong military ally. I think it is likely that Germany will focus on specific nanoweapons, such as mini-nukes and nanobots, while placing high emphasis on continuing to market its consumer, industrial, and medical nanotechnology products throughout the world. Germany will continue as a member of the European Union and participate in the nanotechnology development thrusts of European Framework Programmes for Research and Technological Development. Germany's importance as a NATO nation will grow as it becomes clear that it is a nanoweapons power.

Russia will be a nanoweapons power but not a superpower. Russia will continue to view itself as having parity with the United States and China, but that will be self-delusion. Corruption and mismanagement will take its toll on Russia. Russia's oil wealth and close alliance with China allows maintaining only early generations of nanoweapons. All efforts to commercialize nanotechnology will still leave Russia far behind market leaders, like the United States, Germany, and Japan. Russia will not have a Singularity Computer. Russia will maintain its nuclear arsenal, but it no longer will offer a strategic offensive threat. Singularity Computers in highly advanced technological countries will develop effective countermeasures to nullify any Russian use of nuclear missiles against them.

Saudi Arabia will become a nanoweapons power, using its oil as a strategic bargaining chip to gain computer and nanoweapons technology from the United States. It will also use its oil wealth to buy world-class nanotechnologists and computer technologies. I judge that Saudi Arabia will seek to become a NATO nation. It would be in the best interests of Europe and the United States to have another capable NATO ally in the Middle East.

- Japan will not become a nanoweapons power but will remain a nanoweapons capable nation. Japan will have singularity computer technology but will not build a Singularity Computer. Japan will continue to focus its nanotechnology on commercial, industrial, and medical applications. Japan will remain a "protectorate" of the United States and an important trade partner. Given its lack of military prowess, I believe the United States will not upgrade Japan's relationship to "ally."

Disruption 2: The new cold war will cause every nation to live in fear, even the superpowers. The doctrine of mutually assured destruction (MAD) will not be a clear deterrent.

Disruption 3: The new cold war will give rise to new military alliances. I expect Russia and China to form a military alliance. Their alliance will be in response to the new NATO alliance, which now includes nations with Singularity Computers and nanoweapons. I also think it likely that China will force North Korea to abandon its military aspirations and become a protectorate of China. My rationale is that China will harbor great concerns regarding North Korea's unpredictable military behavior. I also think most nations of the world will pressure China, through trade sanctions, to disarm North Korea.

In the atmosphere of the new cold war, major world religions will unite and proclaim nanoweapons to be inhumane and a clear violation of the natural order. Some philosophers and scientists will openly warn of the increased danger of Singularity Computers combined with nanoweapons. They will argue that each separately presents a 5 percent risk of human extinction by the year 2100, but combined, the risk is much greater than 10 percent, since one technology empowers the other. This is similar to using matches near TNT. Matches represent a risk of fire. TNT represents a risk of explosion. Using them in close proximity edges the combined risk of a catastrophe to a near certainty. Faced with this reality, the United Nations will seek to broker a new arms reduction focused on nanoweapons.

Some superpowers may believe the world is safer than it was during the first Cold War. Unlike nuclear weapons, nanoweapons offer options, from assassinating one person to being a weapon of mass destruction. From my point of view, this mind-set increases the probability of using nanoweapons. The issue is that one use of nanoweapons sets a precedent and may lead to an escalation.

Those nations without Singularity Computers and nanoweapons will seek to:

• Form a military alliance with a superpower. In this regard, I see NATO expanding its role, becoming stronger, and accepting new members. A new "Warsaw Pact" may emerge, with Russia and China forming its core. It will be tough for nations without nuclear or nanoweapons to go it alone.

• Have nuclear weapons as some form of deterrent. Even if they detonate the nuclear weapons within their own borders, it would threaten a nuclear winter that could render the Earth uninhabitable. According to a 2014 report published in the journal *Earth's Future*, out of the more than 15,000 nuclear weapons in existence, the detonation of even 100 relatively small 15 kiloton bombs (each bomb equal to the atom bomb dropped on Hiroshima) would eject over 5.5 million tons of black soot into the atmosphere. This soot would block sunlight, causing global temperatures to drop for twenty-five years. It would also temporarily destroy a significant portion of the Earth's ozone layer, which would allow an 80 percent increase in UV radiation to reach the Earth's surface. The result would be the destruction of the Earth's ecosystem. The entire concept of a nation threatening to detonate its own nuclear arsenal would be an attempt at establishing a new doctrine to replace MAD. In this case, they would threaten to render humanity extinct. Thus the new doctrine would likely be termed "mutually assure extinction." To accomplish this, a nation could set up their nuclear weapons to autonomously detonate unless someone enters a safe code at predetermined intervals. This is a form of a dead man's switch.

For example, the engineer of a train must continually engage the throttle. If something were to happen to the engineer, the train would come to a stop. The new cold war may force non-NATO nuclear powers like India and Pakistan into this position, essentially threating worldwide nuclear suicide.

I also expect to see some of the same elements that characterized the first Cold War to be prominent in the new cold war, such as:

• Propaganda—each side will engage in portraying themselves as victims in an attempt to sway world opinion to their side. I don't think there will be as much saber rattling, since each side will seek to keep their nanoweapons secret and not provoke an undesirable response, such as trade sanctions.

• An arms race—a nanoweapons arms race will rage, similar to the nuclear arms race during the first Cold War. It will be difficult to quantify costs, since Singularity Computers will design and build the nanoweapons. I suspect the cost will be in lost opportunities to use the computers for the further benefit of humanity.

• Espionage—many nations, large and small, will engage in espionage, especially with regard to computer technology and nanoweapons. Spy nanobots will replace humans, but satellite surveillance will continue.

• Clandestine operations—nations will continue to use special forces and political rebels to destabilize adversaries' governments.

• Economic wars—nations will trade with governments deemed friendly and place sanctions on governments deemed adversarial.

• There will also be additional elements to the new cold war:

• A superintelligence race—superpowers will seek to have the most advanced computers, which will lead to an intelligence explosion. Each generation of Singularity Computers will design another one even more powerful. Many AI authorities warn that this may ultimately lead to human extinction.

• A redefinition of a nation's monetary value—when it becomes obvious that Singularity Computers are the key to a nation's prosperity and military prowess, the new currency will be energy and specific raw materials. The Singularity Computers will require energy and raw materials to function, build factories, enable prosperity, and build military prowess. This is the new reality. Scientifically speaking, Einstein taught us that energy and mass are equivalent in his iconic equation $E = mc^2$, where E is energy, m is mass, and c is the speed of light in a vacuum. Eventually it will all boil down to energy. With it, we will be able to make any material we need. Where do we get that kind of energy? This is speculation, but perhaps in the twenty-second century we will learn to harness the complete energy of a star, like our Sun. I do not want to digress further, so for now and the latter half of the twenty-first century, consider a nation's monetary value to equate with energy (i.e., fossil fuels, nuclear, solar, and wind) and raw materials.

Each superpower will also grapple with a new issue, namely, preventing a Singularity Computer from gaining too much military prowess. Let us frame this new issue. A Singularity Computer, essentially a computer more intelligent than the entire human race, may feel threatened by humanity. Its database will clearly show that humanity is an unpredictable, emotional, warring species, with the capability to destroy Earth and therefore to destroy it. Today many researchers argue that a Singularity Computer will represent a new life-form, artificial life, and as with all life-forms, it will seek to protect itself. While I tend to agree, I do not think it necessary for it to be a life-form to seek to protect itself. I judge its logic alone will lead to that. Let me illustrate this with a real example.

Recent experiments, performed in 2009 at the Laboratory of Intelligent Systems in the Swiss Federal Institute of Technology in Lausanne, suggest even primitive artificially intelligent machines are capable of learning deceit, greed, and self-preservation without being programmed to do so. For insight, let us examine the experiments. A research team programmed small, wheeled robots to find "food" (signified by a light colored ring on the floor) and avoid "poison" (a dark-colored ring). Each robot received points the longer it stayed close to food and lost points when it was close to poison. The researchers programmed each robot to flash a blue light when it found food that the other robots could detect. In effect, the researcher programmed the robots to work together to find food. "Over the first few generations, robots quickly evolved to successfully locate the food while flashing the blue light. This resulted in a high intensity of light near food, which provided social information allowing other robots to more rapidly find the food," according to the authors. Following each experiment, the research team would use the data to "evolve" new generations of robots, copying the artificial neural networks of the most successful robots to lesser successful robots. The space around the food, a light colored ring on the floor, was limited. If a robot found food and flashed its light, the other robots moved in, bumping and jostling each other. Even the original robot that found the food could end up being jostled out. By the fiftieth generation, some robots that found food learned not to flash their lights. In effect, some robots became deceitful and greedy. It also implies they learned self-preservation. After several hundred generations, almost all the robots stopped flashing their lights when they found food. Even more startling, some robots flashed their lights to mislead the other robots.

These robots were not self-evolving (able to learn through experience). The researchers had to help them evolve. However, what happens when robots have intelligence equal to or greater than humans? The Lausanne experiment suggests they will act in their own best interest, even ignoring their programming.

The concern regarding Singularity Computers gaining too much military prowess is that they could use their military might on us. If they "reason" that we, humanity, are a threat to their existence, they will seek to eliminate the threat. In such a situation, our own weapons would turn against us.

A related concern is allowing Singularity Computers to interface with one another. Imagine a network of Singularity Computers. Potentially they could pool their intelligence and devise an ingenious plan to gain control over us or even eliminate us. The renowned science fiction author Isaac Asimov anticipated this threat and created the Three Laws of Robotics. He introduced them in his 1942 short story "Runaround."

1. A robot may not injure a human being or, through inaction, allow a human being to come to harm.

2. A robot must obey the orders given it by human beings except in cases where such orders would conflict with the First Law.

3. A robot must protect its own existence as long as such protection does not conflict with the First or Second Laws.

While these laws appear to address the issue in principle, the Lausanne experiment suggests that they would need to be hardwired, not expressed in software.

Let us briefly summarize the world in the year 2050 as envisioned in this chapter:

• Superpowers are defined by the prowess of their Singularity Computers and nanoweapons.

• Singularity Computers offer humanity the potential to end human suffering, extend human longevity, and increase human intelligence, but they also threaten human extinction.

• Nanoweapons offer unparalleled options, from surveillance to destruction, from assassination to mass annihilation, but controlling them is problematic, and loss of control could result in human extinction.

• A new cold war has the entire world in its grip.

• The new cold war fuels a new arms race, with each super-power seeking to have the best Singularity Computers and nanoweapons.

• Combining Singularity Computers with nanoweapons increases the threat of human extinction higher than 10 percent by the year 2100.

• Nuclear powers, without Singularity Computers and nanoweapons, adopt a mutually assured extinction strategy to avoid being attacked by nanoweapons.

• The "dial the level of destruction" aspect of nanoweapons makes their use in conflict greater than the use of nuclear weapons.

The United States and Russia are already deploying nanoweapons, which offer more options than conventional weapons. For example, the United States' new laser weapon can disable an aircraft without destroying it. Given our history as a warrior species and the options that nanoweapons offer, it is a near certainty that nanoweapons will play a role in future warfare. The only question is, Will humanity survive the nano wars?

11 The Nano Wars

> If men can develop weapons that are so terrifying as to make the
> thought of global war include almost a sentence for suicide, you would
> think that man's intelligence and his comprehension . . . would include
> also his ability to find a peaceful solution.
>
> —DWIGHT D. EISENHOWER

With the challenging international landscape today, the average
person and even some military commanders may not be able to
define war, let alone the "spectrum of conflict." In 2006 the pre-
eminent Oxford military historian Hew Strachan made this point
in a lecture: "If we are to identify whether war is changing, and—if
it is—how those changes affect international relations, we need to
know first what war is. One of the central challenges confronting
international relations today is that we do not really know what is
a war and what is not. The consequences of our confusion would
seem absurd, were they not so profoundly dangerous."

This may seem like a philosophical splitting of hairs, but it is
not. The spectrum of conflict defines our level of involvement
and the types of weapons used in each component of the spectrum, including the use of nanoweapons. Therefore, before we
can discuss nano war, we need to define war.

The first crucial thing to understand is that in contemporary
times there is no peace, if we define peace as a period absent
of conflict. Most Western cultures think of war in binary terms,
namely, war or peace. For the most part, they cling to Prussian

general Carl von Clausewitz's classic nineteenth-century definition: "War therefore is an act of violence to compel our opponent to fulfil our will." Our experiences in two world wars served to reinforce our thinking, along with our concept that advanced weapons, like the atomic bomb, are fundamental to winning such wars. Military strategists categorize those wars as "large theater conflicts."

Our adversaries do not share our binary perception of war, and there is strong evidence that they are right. We find ourselves in a spectrum of conflicts, including:

- Limited conventional wars
- Counterinsurgency and stability operations
- Hybrid wars
- Ambiguous wars
- Gray zone wars

Limited Conventional Wars

Limited conventional wars are conflicts between nations using conventional military means, limited by geographic boundaries, types of targets, and disciplined use of force. Most notable is the absence of nuclear weapons. Examples include U.S. involvement in Operation Desert Storm (1991) and the U.S. invasion of Iraq (2003).

In both examples, the United States used conventional weapons against military targets while attempting to avoid civilian casualties. The conflict was limited in geographic scope. This represented a highly disciplined use of force.

Counterinsurgency and Stability Operations

After Operation Desert Storm, the U.S. public and military elite began to believe three questionable paradigms:

1. War is tidy and well defined.
2. War can be won quickly with low casualties.

3. War is won with advanced military weapons and well-trained troops.

Following our invasion of Iraq in 2003, which has led to our continued involvement in the Middle East to maintain security, most would agree those paradigms are false. Instead of winning peace, we found it necessary to engage in counterinsurgency and stability operations.

We should have remembered our own national history. If the three paradigms were true, we would still be a British colony. It appears that many of our military leaders have forgotten the lessons learned during the American Revolution, 1775–83. Thirteen small colonies defeated one of the greatest military powers in existence, Great Britain. How did we win? Let us look at the salient aspects.

• First, the conflict became larger than just the original 20,000 Patriot militia fighting the 50,000 British regulars, 30,000 German (Hessian) mercenaries, and 50,000 Loyalists, colonists that remained loyal to Great Britain. The conflict evolved into a global war, during which the Americans formed alliances with the French, Spanish, and Dutch to fight the British.

• Second, the British largely distrusted the Loyalists, alienating them and effectively reducing their forces by over a third. British concepts about war and conducting military operations ultimately led to defeat.

• Third, the Continental army and militias used guerrilla tactics learned during their battles with the Indians. General Washington's strategy was to avoid large-scale confrontations with the well-equipped royal army. This made it impossible for the British to win a decisive victory. In fact, only once during the entire war did an American army surrender to British forces. That one victory in 1780, at Charleston, South Carolina, hardly constituted a knockout blow to the Americans. A skirmish typically would involve militia behind trees shooting British troops.

After briefly attacking, the militia would disappear into the safety of the forest. In addition, since the militia wore civilian dress, the British could not distinguish militia from Loyalists.

• Fourth, the Americans' alliance with France, Spain, and the Dutch served them well. France aligned itself with the colonists in 1775 by secretly sending them money and weapons. In 1778 France declared war on Brittan and increased its involvement by sending an army to fight with the Americans. In 1776 Spain began funding Roderigue Hortalez and Company, which provided military supplies to the revolutionaries. In 1781 Spain also funded the siege of Yorktown. In 1782 the Dutch was the second European nation to diplomatically recognize the Continental Congress, or as we term it today, the United States. (In 1777 Morocco became the first country to diplomatically recognize the United States as an independent nation.) This made the American Revolution a global confrontation in which Great Brittan had no allies. Most historians doubt the Americans would have prevailed without the help of its allies.

• Fifth, the British focused on capturing cities. In fact, they did capture New York City and held it for the duration of the war. Capturing cities was not difficult for the British, but this proved a questionable strategy. The colonists did not have a national capital that the British could capture and thereby end the conflict. In addition, without the support of Loyalists, the British found it difficult to hold captured cities. For example, the Continental army drove the British out of Boston in 1776.

• Sixth, the geographical size of the colonies made it impossible for the royal forces to end the rebellion via occupation. Although there were four cities, Philadelphia, New York, Boston, and Charleston, 90 percent of the colonists lived in rural areas, which includes Delaware, Pennsylvania, Massachusetts Bay Colony (which included Maine), New Jersey, Georgia, Connecticut, Maryland, South Carolina, New Hampshire, Virginia,

New York, North Carolina, Rhode Island, and Providence Plantations. New York State alone is almost 55,000 square miles. The royal forces numbered about 80,000, not counting the Loyalists. This means the royal forces could not station even one soldier per square mile occupied by the colonists.

• Widespread support for the Revolution cut across regions, religions, and social ranks. Thousands of farmers, artisans, ex-servants, slaves, shopkeepers, merchants, and laborers took up arms to fight the British. Today's military strategists would characterize the tactics of the Americans as "unconventional asymmetrical warfare."

My point in capsulizing the salient aspects of the Revolution is that our commonly held ideas regarding the nature of war are largely erroneous. War is not tidy and well defined. Superior weapons and well-trained troops do not guarantee victory.

Hybrid Wars

One level down from limited conventional war is hybrid warfare, which includes "irregular wars," or any blend of regular and irregular tactics. According to the 2008 U.S. Army Posture Statement, the United States defines irregular warfare as "a violent struggle among state and non-state actors for legitimacy and influence over the relevant populations." It embodies indirect and asymmetric approaches that avoid direct confrontations with strong forces. The 9/11 terrorist attacks on the United States would qualify as an example.

Ambiguous Wars

According to the Center for Naval Analysis, the phrase "ambiguous warfare" is a term that "applies in situations in which a state or non-state belligerent actor deploys troops and proxies in a deceptive and confusing manner—with the intent of achieving political and military effects while obscuring the belligerent's direct participation." An example is Russia's seizure of Crimea in 2014, ordered by Russian president Vladimir Putin to protect "Russian citizens and compatriots on Ukrainian territory." Moscow's actions deliberately obscured the annexation of Crimea and delayed Western responses. The initial Russia troops who seized Crimea wore uniforms without national insignias. Was it a Russian invasion? Was it a move to protect Russian citizens? You be the judge.

Gray Zone Wars

We can define gray zone wars as deliberate activities by a nation, or fractions within a nation, that seek to achieve strategic goals without the use of military forces. China's assertive behaviors in the South China Sea, such as the deliberate use of security forces to assert sovereignty in and around contested shoals and islands in the Pacific, is an example of a gray zone conflict. Their behavior aims at changing the norms of international behavior and asserting their interpretation. Russia is also engaging in gray zone conflicts, such as seeking to extend its influence in numerous regions without triggering an armed NATO response.

Defense policy veteran Nadia Schadlow made this observation regarding the U.S. reaction to gray zone conflict: "By failing to understand that the space between war and peace is not an empty one—but a landscape churning with political, economic, and security competitions that require constant attention—American foreign policy risks being reduced to a reactive and tactical emphasis on the military instrument by default."

I liken this to someone who is rude and behaves just short of provoking an altercation. For example, a neighbor may use a portion of your property to have an outdoor party without asking permission. The neighbor may even invite you to the party. In such a situation, many people would be inclined to let it go without making an issue. First, they may want to stay on friendly terms with their neighbor. Second, it is temporary. The trespassing lasts only for the duration of the party. Third, the neighbor invited you to join the party. The neighbor's behavior is rude, since he or she never sought your permission, but it is unlikely to cause an altercation or even an exchange of words. In real life, this happens all the time. You may even attend the party and bring a case of beer as a goodwill gesture. However, now the neighbor has established a precedent, namely, it is okay to use your property for a party. Consider how difficult it will be to object to its use next time. That is the exact nature of gray zone conflict, except it occurs on the national and international levels.

Nano War and Its Place in the Spectrum of Conflicts

Now that we have discussed the spectrum of conflicts, you may ask, Where does a nano war fit into the spectrum? To address this question, let us start by reiterating two important points:

1. Nanotechnology is an enabling technology. Think of it as a category of technologies, not a specific technology.

2. Nanoweapons are weapons that exploit the use of nanotechnology.

We can define a nano war as any war that includes the use of nanoweapons. To categorize nano wars in the spectrum of conflicts, it is important to define the categories of nanoweapons.

Let's define the categories, discuss each, and give some concrete examples of the nanoweapons in each category.

1. Passive nanoweapons: This category relates to any use of nanotechnology in warfare that has a nonoffensive/nondefensive application, but may increase the effectiveness of conventional or strategic weapons. In many cases, passive nanoweapons will have a commercial, industrial, or medical counterpart. Representative examples include:

a. Nanoelectronic integrated circuits: These advanced integrated circuits form the backbone of Department of Defense computers and smart weapons. While nanoelectronic integrated circuits themselves are not a weapon in the traditional sense, they enable a wide spectrum of conventional and strategic nanoweapons.

b. Nanoparticles: All branches of the U.S. military will use nanomedicine, such as nanosilver-impregnated bandages to inhibit infection. The Navy is using nano coatings to inhibit corrosion and marine growth on its ships. The Army is using nano-enhanced materials to develop lighter, stronger armor for combat troops. Nanotechnology-based explosives can increase the destructive power of conventional explosives.

c. Nanosensors: These sensors are finding wide application in every branch of the U.S. military because they offer unparalleled opportunities to interact (i.e., sense) at the molecular level. This makes them extremely effective as biosensors and chemical sensors. They are also finding applications in image sensing.

d. Nanotechnology-based commercial and industrial products: These are finding military applications, including enhanced steel and concrete to build command and control centers, armor plating, and stronger artillery.

2. Offensive tactical nanoweapons: This category relates to offensive weapons whose nanotechnology components enhance their tactical capabilities.

a. Nanotechnology-based lasers: The Navy's is deploying laser weapons that offer options to disable or destroy adversary targets. The Army is fielding infantry lasers, mounted to conventional vehicles, that are able to fire laser bursts of energy to destroy cruise missiles, artillery, rockets, and mortar rounds.

b. Superior snipers: The Army is working to make their snipers completely invisible under a nanomaterial cloak of invisibility with the ability to fire smart bullets that make a one-mile kill routine.

c. Smart artillery: The Army is developing smart artillery, which may profoundly change the role of artillery or even the role of the Army, as a land force, to a strategic force capable of engaging air, sea, and land targets.

d. Mini-nukes: These nuclear weapons are arguably conventional, since they do not result in residual radiation or radioactive fallout. As such, they will likely find offensive applications in every branch of the U.S. military

e. Nanorobotics: Every branch of the U.S. military will use this class of robotics in offensive applications. The Army is developing the Autonomous Rotorcraft Sniper System, an unmanned helicopter carrying a remotely controlled sniper rifle. The Navy is testing an offensive unmanned drone vessel, the Anti-Submarine Warfare Continuous Trail Vessel, to track enemy submarines and limit their tactical capacity for surprise. Navy drone vessels will play a role similar to their Air Force counterparts. The Air Force is developing bird-sized and insect-sized drones that can carry an explosive payload, sufficient to kill a single adversary or destroy an adversary's command and control installation.

3. Defensive tactical nanoweapons: This category relates to defensive weapons whose nanotechnology components enhance their tactical capabilities. Representative examples include:

a. Nanorobotics: Every branch of the U.S. military will use this class of robotics in defensive applications. The Army, for example, will use it to detect and disarm improvised explosive devices (IEDS). The Air Force will use it to develop surveillance drones the size of birds and insects. The Navy is already using it in their swarm boats.

b. Nano-enhanced metal: New steels, with ten times the strength of normal steel, will find wide military application, from Army armored troop carriers to Navy ships. The nano-coated metals may become a standard "mil-spec" for all metal components, due to the improved strength and corrosion resistance. Being lighter, Army vehicles and Navy vessels could travel farther and faster with the same amount of fuel they now consume, with coated engine parts that virtually never wear out.

c. The invisibility cloak: This technology is likely to find numerous applications in all branches of the military. In addition to providing a new level of cover for Army snipers, the Army will use it to add stealth to its soldiers and weapons. Future drones may not only have low radar signatures but also be optically invisible. Imagine, for example, a fly-sized invisible drone. It could sit on the table of an adversary's command and control center to provide real-time military intelligence on the adversary's plans

4. Offensive strategic nanoweapons: This category relates to weapons whose nanotechnology components enhance their strategic capabilities. It also includes offensive autonomous smart nanobots. Representative examples include:

a. Autonomous smart nanobots—these would represent the ultimate weapons of mass destruction. A nation, or even a terrorist organization, could program them to attack an adversary's infrastructure, its leaders, and its people. Current technology already confirms that making such nanobots is possible. Extrapolation of current technology to the year 2050 suggests that building self-replicating smart nanobots (ssn) will be possible. The issue will be maintaining control of these nanoweapons, which represent the ultimate doomsday devices and threaten human extinction.

b. Hypersonic glide missiles: These will likely require nanomaterials, nano-enhanced fuels, and nanoelectronic guidance systems. They will be capable of reaching a target halfway around the globe in an hour. The United States claims that they will only use conventional warheads on such missiles, but they clearly have the capability to deliver nuclear weapons and nanoweapons. Currently the United States is the clear leader in hypersonic glide missile technology. As of this writing, there is no known defense against such missiles.

5. Defensive strategic nanoweapons: This category relates to defensive strategic weapons whose nanotechnology components enhance their strategic capabilities. It also includes defensive autonomous smart nanobots. Representative examples include:

a. Autonomous smart nanobots: Nations could program these nanobots and the future self-replicating smart nanobots to destroy a nation's offensive weapons, including nanoweapons. The issue, again, is controlling these nanoweapons to avoid threatening human extinction.

b. Antiballistic missile defense system: The United States has limited antiballistic missile defense capabilities, but I judge advancements in nanoelectronics and solid, stable nano-enhanced fuels for ballistic missiles will change that position. Based on the rate that nanotechnology and artificial intelligence are advancing, I project that within a decade, the United States will have the capability to destroy adversary ballistic missiles in flight.

Now that we have categorized nanoweapons, let us discuss their application in warfare. Since nanoweapons offer numerous options, from killing one person to wiping out a nation, from destroying civilian infrastructure to destroying military hardware, their use in conflict is certain. Indeed, we have already discussed several nanoweapons, such as the Navy laser and nanoparticle-enhanced explosives, which are combat-ready. Numerous others are likely to be used in future conflicts. The most likely are nanoweapons in the following categories:

• Passive nanoweapons
• Offensive tactical nanoweapons

• Defensive tactical nanoweapons

In my opinion, nations will use passive and offensive/defensive tactical nanoweapons in conflict. This opinion reflects the historical trend of warfare. Since the emergence of our species, each war has given rise to weapons of greater destruction. Deployment and use of passive and tactical nanoweapons will just continue a trend that has gone on for over 200,000 years. Given this opinion, it is natural to ask, What about the deployment and use of strategic nanoweapons?

Even in warfare, humanity draws a line when certain weapons threaten the extinction of our species. Humanity is not a perfect species, but we collectively act to prevent our own extinction. I think of it as a collective self-defense reaction. There are several examples that illustrate this:

1. Treaty on the Non-Proliferation of Nuclear Weapons
The Cold War tensions between the United States and the Soviet Union brought the world to the edge of a nuclear confrontation. This resulted in a widely held belief among nations that proliferation of nuclear weapons would reduce world security, increase the risks of miscalculation and accidents, and eventually lead to a nuclear conflict. Historically, the United Nations opened the treaty for signature in 1968. It went into force in 1970 and was "extended indefinitely" in 1995. Today 191 nations are signatories of the treaty, but five nations are not, including India, Pakistan, and North Korea, which have openly tested and declared that they possess nuclear weapons. Israel is also a nonsignatory and refuses to disclose its nuclear weapons capabilities. South Sudan never joined the treaty, but there is no evidence they possess nuclear weapons. Thus, out of the 196 world nations, only 9 have nuclear weapons, and 5 of the 9 are signatories of the treaty. Although this is not ideal, more countries have adhered to this treaty than to any other arms limitation and disarmament agreement. No one, not even nonsignatories, has used nuclear weapons in conflict since World War II. Historically,

since 1945, nations have engaged in 250 major conflicts killing over 50 million people. This provides evidence that humanity is able to restrain itself. However, this restraint came with a price. The world witnessed the detonation of two atomic bombs used in World War II, killing over 250,000 Japanese, and 520 atmospheric nuclear test detonations, including automorphic testing of the largest nuclear weapon, the 50-megaton Russian AN602 hydrogen bomb. Many experts believe these nuclear detonations and the associated radioactive fallout are still causing an increase in cancer-related illnesses. My point is that people of the world knew that nuclear weapons presented an unparalleled risk to humanity's survival. Because people were informed, there was a grassroots thrust to limit the number of nuclear powers, which influenced world leaders to commit their nations to the treaty.

2. Limited Test Ban Treaty

The Limited Test Ban Treaty (LTBT) is officially the treaty banning nuclear weapon tests in the atmosphere, in outer space, and under water. The Soviet Union, the United Kingdom, and the United States signed and ratified this treaty in 1963. Most countries, with the exception of China, France, and North Korea, have signed and ratified the treaty. Even nonsignatory countries are abiding by it. There have been several violations, but to a large extent the treaty slowed the nuclear arms race and addressed rising concerns about radioactive fallout as a result of nuclear weapons testing. At the time of the treaty's ratification, most of the world's population knew that nuclear fallout could be just as deadly as the initial blast. Again, people were informed, and their leaders acted in accordance with national sentiments.

3. Biological Weapons Convention (BWC)

Officially this is the Convention on the Prohibition of the Development, Production, and Stockpiling of Bacteriological (Biological) and Toxin Weapons and on Their Destruction. The BWC took the 1925 Geneva Protocol, which prohibits use but not possession or development of chemical and biological weapons, to the next step. It represents the first multilateral disarmament treaty banning the production of an entire category of weapons. It went into force in 1975, and 172 nations are signatories. Again, the use of chemical weapons in World War I and the 1918 flu pandemic, which afflicted one-third of the world's population, killing an estimated 50 million people, was sufficient to convince the world population and their national leaders to adopt the BWC. (However, although the BWC is generally successful, there have been some violations. The most notable exception is Iraq's 1988 use of chemical weapons against the Kurdish people in the city of Halabja forty-eight hours after the city fell to the Iranian army and Kurdish guerrillas. In 2010 the Iraqi High Criminal Court condemned the attack as an act of genocide and a crime against humanity.

Conclusions

My main point in presenting these three examples is that their adoption required an informed population. This point, along with the previous discussions, allows us to draw conclusions regarding the development, deployment, and use of strategic offensive/defensive nanoweapons.

> Conclusion 1: Once presented with evidence that a class of weapons can threaten human extinction, humanity will act to protect itself. This will be true of strategic offensive/defensive nanoweapons. Unfortunately, most of the world's population is oblivious to the existence of nanoweapons. They garner no headlines or primetime news coverage. There are no treaties adopted to prevent nations from developing and deploying them. This suggests that strategic offensive/defensive nanoweapons will be developed, deployed, and used in conflict before the world understands the threat that they pose to our existence.

Unfortunately, even deploying strategic nanoweapons represents a risk, which is greater than deploying nuclear weapons. I judge the risk to be greater because strategic nanoweapons will embody artificial intelligence, and we risk losing control over them and the Singularity Computers that built them.

Conclusion 2: The threat of a full-out nano war will loom large in the second half of the twenty-first century. This threat is large because identifying the source of a nanoweapons attack will be problematic due to the inherent stealth associated with them. As a result, nations and terrorist groups will find it compelling to unleash offensive tactical nanoweapons on an adversary. However, any attack increases the possibility of a full-out nano war, which like a full-out nuclear war would threaten humanity's survival.

The new cold war of the 2050s and beyond will be extremely dangerous, even more so than the first Cold War. The world knew what was at stake regarding nuclear weapons. This knowledge forced nations to dial back the threat of nuclear war. This is not the situation with nanoweapons. Currently the bulk of the world population does not even know that nanoweapons exist, and major powers are in a new nanoweapons arms race. This argues that the next cold war has a greater probability of escalating into a strategic nano war, which would push humanity to the brink.

12 Humanity on the Brink

All of us might wish at times that we lived in a more tranquil world, but we don't. And if our times are difficult and perplexing, so are they challenging and filled with opportunity.

—ROBERT F. KENNEDY

Many great tragedies require a series of events to take place, to line up like dominos where the fall of one domino ultimately results in the fall of all. Each event may appear innocuous at the time, like one falling domino. In and of itself, it may engender no feelings of danger or concern. It is only after the series of events results in some unforeseen tragedy that we realize the significance of each event. History is ripe with examples. Let us consider one, the nuclear disaster at Chernobyl.

The Chernobyl Nuclear Disaster

April 25, 1986, was an important day for day shift workers at the Chernobyl nuclear power plant. After weeks of planning, it was time for them to participate in an experiment scheduled to test a potential safety emergency core-cooling feature of the Chernobyl #4 nuclear reactor. In addition, a team of electrical engineers was present to test the new voltage regulating system. The experimental procedure was well defined:

1. Run the reactor at a low power level, between 700 MW and 800 MW. (MW stands for megawatts, 1 megawatt equals 1,000,000 watts. Thus a plant running at 700 MW produces 700 mega-

watts per hour, which is sufficient to run 7,000,000 100 watt lightbulbs for one hour.)

2. Run the steam-turbine generator at full speed.

3. Shut off the steam to the turbine generator as soon as conditions 1 and 2 are achieved.

4. Record the turbine generator performance to determine whether it could provide the bridging power for coolant pumps until the emergency diesel generators start to power the cooling pumps automatically. (The engineers knew that it would take the emergency diesel generators about one minute to start and power the cooling pumps. The plan was to use the residual power in the turbine generator to bridge the one-minute gap.)

5. Freewheel down (i.e., allow to run down in its normal course) the turbine generator after the emergency generators reach normal operating speed and voltage.

The objective of the test was to ensure that the residual power in the turbine generator would supply the power necessary to keep the main water-cooling pump operational until the emergency generators came online.

On paper, the test appeared straightforward and necessary to ensure a viable backup in the event of a power failure. But then a series of events turned the test into a disaster. Appendix III details these events for those who seek a deeper understanding. Here, I will only present my observations:

1. The experimental procedure lacked contingency planning. After reviewing the series of events, it is apparent that no what-if exercises were discussed. This required that every element of the Chernobyl experiment result in the expected outcome. In my experience with state-of-the-art development and manufacture of integrated circuits and sensors, I can state it is extremely rare in a complex experiment for every result to be anticipated or for the experiment itself to proceed per plan.

2. External events, such as another power station going off grid , interrupted and delayed the experimental procedure. An event like this would have been anticipated in a properly performed what-if exercise.

3. The delays caused by another power station going off grid ultimately left an inexperienced night shift engineer in charge of control room personnel who had little training to conduct the experiment. In a perfect world, this event alone would result in suspending the experiment.

4. When events deviated from the experimental procedure, the lead engineer and control room personnel did not suspend the experiment, but continued it under dangerous circumstances. This illustrates the inexperience and disregard for safety that characterized the participants.

5. Poor judgment, likely the result of stress and inexperience, played a significant role in turning the experiment into a disaster. The series of events illustrates key times that automated safety procedures were disengaged and warning alarms were intentionally ignored.

We know that the explosion of Chernobyl's reactor #4 resulted in a fireball, smoke, and steam that carried radioactive material into the atmosphere. We also know that the Soviet government did not immediately warn its people or their neighboring nations. Even though thirty-two people died and dozens more suffered radiation burns in the first few days of the crisis, the Soviet government only admitted the accident after Swedish authorities, more than eight hundred miles to the northwest of Chernobyl, reported radioactive fallout. An estimated 985,000 died, mostly from cancer, due to the Chernobyl accident. It is widely recognized as humanity's worst nuclear disaster.

The Chernobyl explosion illustrates these lesson points:

1. Modern technology, such as nuclear technology and nanoweapons, can be complex and require high expertise to deploy and control.

2. Humans have limitations, and the complexity of modern technology challenges those limitations.

   a. Our ability to anticipate and control the outcome of a long series of complex technological events is questionable. The Chernobyl disaster presented the lead engineer and control room personnel with just that situation, and they failed to recognize, in a timely manner, the danger they faced. Ultimately, each event appeared to them as isolated, like the fall of a single domino. They failed to understand the ultimate safety hazard embodied in the series of events. In effect, they did not recognize that the fall of one domino would ultimately result in the fall of all dominos.

   b. Human judgment can be severely impaired by lack of experience and stress.

3. Control of nuclear technology and nanoweapons will always present risks, including risks to humanity's survival.

A World with Nanoweapons Superpowers

We will use the lesson learned from the Chernobyl disaster by applying them to nanoweapons. To do this, let us fast forward to the second half of the twenty-first century. If the world remains on its current trajectory, the second half of the twenty-first century will be similar to the Cold War in the second half of the twentieth century, where nuclear superpowers could determine the world's fate with the push of a button. However, this time the superpowers will have added tactical and strategic nanoweapons to their arsenal.

If this view of the world is correct, what will characterize the world environment beyond the year 2050?

1. First, we will have all the cold war elements we discussed in chapter 10, including military alliances, worldwide nuclear

suicide threats, propaganda, a nanoweapons arms race, espionage, clandestine operations, economic wars, Singularity Computers (computers that are more intelligent than the combined intelligence of humanity), and monetary systems based on raw materials and energy.

2. Second, technology complexity will be orders of magnitude greater than it is today, which suggests:

a. Developing and deploying nanoweapons will be problematic, with many Chernobyl-like disasters in the offering.

b. Controlling Singularity Computers will be problematic, as we run the risk that the super intelligent machines may become adversarial toward humanity.

Even with this environment, a question remains, Will humanity understand that its existence is balanced on a knife edge? The answer will depend on one thing, namely, a nanoweapons superpower using a significant fraction of its capabilities in conflict. Think about it. No one understood the implications of nuclear weapons until their use in conflict. No one understood the implications of "stealth" technology until its use in conflict. At this point, a thick veil of secrecy obscures our view of the destructive power of nanoweapons. Using nanotechnology-enhanced conventional weapons will not be sufficient. For example, if the Navy uses its new nanotechnology-based laser to disable a plane, the laser, not the nanotechnology that enables it, will impress the average person. Let's take an example closer to home. When you turn on your brand-new computer, what impresses you, the computer's performance or the nanotechnology that enables it? Most people do not know that their new computer even incorporates nanotechnology. Based on historical precedent, only a definitive combat application of nanoweapons will reveal their true destructive nature.

You may ask, What will constitute a definitive combat application of nanoweapons? Let us consider two examples, short of full-out nano war.

Imagine, a tense situation forming in the China Sea. U.S. carrier fleets prowl the region. An adversary threatens Japan, a U.S. protectorate. The adversary has its own carrier fleets, including submarines with nuclear missiles and surface vessels with missile launch capabilities. From the standpoint of conventional and nuclear weapons, let us assume there is a complete balance of naval power. In this example, let us assume that the adversary prepares to launch its attack on Japan. The adversary's submarines hover at missile launch depth and begin opening their missile hatches. The adversary's carrier launches its planes, and they are heading for Japan. The United States concludes that within seconds the adversary will rain death and destruction on Japan. The United States has no alternative. It activates the self-replicating smart nanobots that infest every element of the adversary's arsenal. Almost immediately, the adversary's submarines begin to sink, the carrier's planes fail in flight and crash into the ocean, the carrier and missile launch vessels begin to sink. In desperation, the adversary attempts to launch missiles, but the missiles never leave the launch pads. Within a few minutes, the adversary has no naval capabilities left. The battle is over before it begins. Now begins the most difficult part.

The adversary launches a formal protest with the United Nations. The elements of the protest claim:

• The adversary was peacefully performing naval operations in the China Sea.

• The United States conducted a sneak attack using nanoweapons.

• The adversary calls on the United Nations to condemn the attack as an act of war by the United States and enact appropriate sanctions against the United States.

The United States presents its data, recorded on its satellites and naval instruments. Using the data, the United States argues that the adversary was beginning an attack on Japan. As a protectorate of the United States, the adversary provoked the use of nanoweapons.

The world is shocked. Headlines and "experts" saturate all media detailing the destructive power of nanoweapons. Humanity begins asking the hard questions:

What kind of nanoweapons did the United States use?

How destructive are these new nanoweapons?

Are nanoweapons controllable?

Do nanoweapons threaten humanity?

I will not take this further. I think the point is clear. The unequivocal use of nanoweapons in conflict will be a wakeup call for the world population.

Genocide

For this example, consider a nanoweapons superpower, Nation-X, with an incredibly large population. There is significant unrest throughout the population due to food shortages and scarce medical attention. Other superpowers, by comparison, provide an unparalleled standard of living for its citizens, including socialized medical treatment, luxurious housing, and even antiaging therapies. In addition, Nation-X is geographically large. A significant portion of its population remains poorly educated and continues to rely on family farming for sustenance. The senior elements of this population continue to suffer normal human illnesses associated with aging. The significant unrest among the general population of Nation-X has given rise to public antigovernment protests. Most protesters feel the government focuses on weapons ahead of caring for its people. A typical protestor's sign reads, "People, Not Weapons." The government of Nation-X is losing popular support, which is eroding its ability to govern.

To regain control, over a three-day period Nation-X takes the following steps:

- Suspends the right of its citizens to protest
- Forces all foreign news agencies to leave the country, creating a media blackout
- Closes its borders
- Shuts down all but government communications

  Phone service is suspended, including land, cell, and satellite service.

  The Internet is restricted to only government websites.

  Email is shut down except for secure email between government officials and government email to other nations.

  It jams any radio or television transmissions that are not government initiated.

The United States and other NATO nation satellites detect large portions of Nation-X's population is dying or already dead. The United States reveals this finding at an emergency meeting of the United Nations.

Nation-X claims they are dealing with a "super influenza" that does not respond to even advanced antiviral pharmaceuticals. It declines all help offered by member nations.

The United States calls for a United Nations confirmation of Nation-X's claim. Nation-X responds that any such inspection would potentially spread the infection and strenuously objects.

The United Nations proceedings drag on for days. Tensions in the world spread. Suppose Nation-X is telling the truth. Will other nations be next? The United Nations' puts a proposed resolution to investigate the super influenza to a vote. The bulk of world nations vote in favor of the resolution, with the provision that the United Nations inspectors remain in Nation X until the world community has a way to deal with this super influenza.

As the United Nations slowly grinds to a resolution, some citizens of Nation-X escape into bordering countries and are given asylum, but to no avail. They too die. The United States obtains the dead bodies of several victims of the super virus and performs autopsies at sea, to ensure that any super virus remains confined to the limits of one vessel.

The United States confirms what they suspected, killer nanobots. Nation-X released smart autonomous killer nanobots on its own people. The nanobots' programming limited their attack to specific geographic regions and specific DNA signatures. With this data, the United States estimated that Nation-X was intent on eliminating about half of its population, with specific attacks on protestors (via DNA signatures), the elderly, and those with terminal illnesses.

The United States brings its data before the United Nations. Nation-X denies any involvement in genocide and argues only the United States has that level of advanced autonomous smart nanobots. The United States is outraged and denies any involvement in the genocide.

The proceedings of the United Nations dominate media coverage. Humanity begins to ask the hard questions:

Did Nation-X commit genocide using advanced nanoweapons?

Would the nanobots remain confined to Nation-X?

I will not take this further. I think the point is clear. The unequivocal use of nanoweapons by a nation, even within its own borders, will be a wakeup call for the world population.

In both examples, I judge humanity will heed the alarm and respond. Similar to humanity's actions in the decades following the use of nuclear weapons in conflict, humanity will seek treaties limiting nanoweapons use and assurances that nanoweapons will remain under control. We are likely to see treaties that parallel those in place for nuclear weapons.

Conclusion

An important question remains, Will humanity's reaction to a nanoweapons event be timely and sufficient to avoid its own extinction?

I believe waiting for history to answer the question is too risky. Nanoweapons, like nuclear and biological weapons, threaten the survival of humanity. Nanoweapons represent the ultimate threat because:

1. Controlling extremely sophisticated nanoweapons may be beyond the capability of human intelligence. Let us recall the Chernobyl disaster presented at the beginning of this chapter. The technology of Chernobyl would be TinkerToys compared to the technology of advanced nanoweapons. To my mind, it is not a matter of if we will have a Chernobyl-like disaster with nanoweapons, but when. The sophisticated nature of advanced nanoweapons makes them another Chernobyl waiting to happen.

2. The Singularity Computers under human direction designed the advanced nanoweapons. However, Singularity Computers represent an inherent threat, in and of themselves. This argues that allowing Singularity Computers to have complete control of advanced nanoweapons could be catastrophic, since this may give them the means to render humanity extinct. Therefore, any nanoweapons designed using Singularity Computers must be hardwired to respond to human control. There must be a fail-safe way for humans to override any Singularity Computer commands to nanoweapons.

With the rise of nanoweapon superpowers and the increased threat of a nano war, nations and the bulk of humanity will recognize the threat that nanoweapons pose to the survival of humanity. It may take a Chernobyl-like accident or the tactical use of nanoweapons to sound the alarm. We humans appear to be a reactive species. It generally requires a sizable event to galvanize us to act.

Given the full capabilities of nanoweapons, nations will become increasingly suspicious of other nations with nanoweapon capabilities, resulting in a higher probability of first-strike warfare and the use of nuclear weapons, as a more controllable alternative to nanoweapons. This is ironic. Today we consider the use of nuclear weapons horrific, given their devastation, residual radiation, and the unpredictable path of radiation fallout. However, the use of nuclear weapons to thwart a nanoweapons attack may represent the best alternative, namely, the lesser of two evils.

In a world where humanity is on the brink of causing its own demise, this chapter makes a case that the ethics and control of nanoweapons will become the #1 United Nations issue. The epilogue talks to this last point and delineates strategies that humanity can employ to avoid falling victim to its own technologies.

Epilogue

The nuclear arms race is like two sworn enemies standing waist deep in gasoline, one with three matches, the other with five.

—CARL SAGAN

I believe that people who have access to current information choose to go through life in one of two ways, namely, either remaining actively informed or remaining purposely ignorant. That may seem harsh and simplistic. However, I see its manifestation in almost every substantive interaction. The information highway we ride on a daily basis overwhelms some people. You probably have heard someone say, "I never watch the news." Do you know people who only read the comics section or only do the crossword puzzles in their local newspaper? I am not being judgmental. I am only observing, and I believe many readers would agree with this observation. In fact, if I have a highly stressful day, I will put aside catching up on the news. Instead, I may listen to music or relax. It is human nature. With today's instant communication, we know about almost every disaster in the world within minutes. I have also observed that most headlines carry bad news. Why is this? In my view, bad news appears to pique more interest than good news. It sells more media advertising. However, lack of information offers no protection against the "slings and arrows of outrageous fortune" that life aims at us. I speculate that as many as half of the nearly million people who died as a result of the Chernobyl disaster did not know that this event had sealed their fate.

Attention to strategic nanoweapons as weapons of mass destruction has been sparse to nonexistent among the media, policy makers, and scholars. In contrast, a vast body of literature exists about nuclear, chemical, and biological weapons. It appears, though, that no connection equates strategic nanoweapons to those weapons of mass destruction, mainly because a cloak of secrecy ensures that nanoweapons remain below the radar. Most people are oblivious to the existence of nanoweapons. This fact, and the threat that nanoweapons pose to humanity's survival, prompted me to write this book.

This book is about providing information on the next and most deadly generation of military weapons the world has ever encountered. The information is concerning. Some of the conclusions drawn from the information are horrific. In many cases, I have presented the facts and left it to the readers to draw their own conclusions. It is important to inform the public. As Thomas Jefferson put it, "If a nation expects to be ignorant and free in a state of civilization, it expects what never was and never will be. An informed citizenry is at the heart of a dynamic democracy."

This book is a compilation of facts about nanoweapons. These facts are hidden in the countless pages online, in books, in articles, and in scientific journals about nanotechnology. For every thousand pages on nanotechnology, there is perhaps one page related to its use as a nanoweapon. You can verify this assertion yourself by simply comparing the Google search returns on "nanotechnology" versus "nanoweapons." Yet, if you are willing to sift through the data, it is possible to extract the facts like gold nuggets from tons of ore. This is what I have done. In addition, I have organized the facts and provided projections. I believe that many readers will use the information to develop their own projections. This is how it should be. Through this process, we learn and refine our understanding.

I know it is a cliché to say, "In no time in history has it been more important than now to have an informed electorate." However, given the challenges that nanoweapons present, the cliché is accurate. It is critical for all to understand the potential threat that nanoweapons pose. We cannot ignore their existence. We cannot ignore the nanoweapons arms race among the United States, China, and Russia. We must address the question, Will it be possible to develop, deploy, and use nanoweapons in warfare without rendering humanity extinct? Let us address this question by examining how humanity historically has dealt with similar weapons of mass destruction.

In previous chapters, I have characterized humanity as a warlike species, but I also believe that humanity learns from history and that we collectively act in self-defense. There is abundant evidence supporting this assertion. Examples include the Treaty on the Non-Proliferation of Nuclear Weapons, the Limited Test Ban Treaty, and the Biological Weapons Convention.

Humanity's existence has historically balanced itself on a knife edge since the invention of weapons of mass destruction (WMD). However, let us be clear on what constitutes WMD. The UN Commission for Conventional Armaments of 1948 defined WMD as including "atomic explosive weapons, radioactive material weapons, lethal chemical and biological weapons, and any weapons developed in the future which have characteristics comparable in destructive effect to those of the atomic bomb or the other weapons mentioned above." So how does this definition affect nanoweapons?

In chapter 11, we divided nanoweapons into five categories:

1. Passive

2. Offensive tactical

3. Defensive tactical

4. Strategic offensive

5. Strategic defensive

If we apply the UN definition, only categories 4 and 5 embody nanoweapons of mass destruction:

• Autonomous smart nanobots (including self-replicating smart nanobots)

- Hypersonic glide missiles

- Antiballistic missile defense systems

This is an important observation. It suggests that there are historical precedents that may provide insight on how to treat strategic offensive/defensive nanoweapons.

Let us begin by examining the most destructive of all nanoweapons: self-replicating smart nanobots (SSN). To my mind, SSN are technological artificial life-forms and analogous to their biological counterparts, viruses. I suggest we consider classifying SSN under the Biological Weapons Convention. This would be an obvious extension of that treaty to cover this new class of nanoweapons.

Typically biological life-forms satisfy five conditions:

1. Are made of cells

2. Obtain and use energy

3. Grow and develop

4. Reproduce

5. Respond and adapt to their environment

Self-replicating smart nanobots would satisfy conditions 2, 4, and 5. They do not have biological cells, and they do not grow and develop. As a result, you may argue they are not, strictly speaking, a life-form. However, I think the conditions we imposed for something to be a life-form may be incorrect, since it relies only on our experience with biological life-forms. Technological life-forms may satisfy a different set of conditions, but still be equivalent to a biological life-form. Science fiction writer Arthur C. Clark said it best: "Whether we are based on carbon or on silicon makes no fundamental difference; we should each be treated with appropriate respect." Clark's comments related to artificial intelligence, but I judge they are extendable to SSN.

I favor extending existing treaties and conventions for weapons of mass destruction to cover strategic nanoweapons. This eliminates the need to "reinvent the wheel," which could take decades and not address the issue in a timely manner.

If we categorize weaponized SSN as within the Biological Weapons Convention, we still have to deal with other strategic nanoweapons, namely, autonomous smart nanobots, hypersonic glide missiles, and antiballistic missile defense systems. Therefore, this one move would not solve the entire problem. The question now becomes: How do we prevent the remainder from threatening human extinction?

I suggest three strategies:

1. Confine hypersonic glide missiles to the delivery of conventional weapons. The United States is already committing to this strategy. Russia is not. In fact, Russia states it will arm its hypersonic glide missiles with nuclear warheads. Obviously, this would make the new cold war extremely dangerous. In my view, Russia is taking this stance because the United States is currently leading in hypersonic glide missile technology, and Russia wants to "one up" the stakes. Unfortunately, this may force the United States to alter its hypersonic glide missiles to carry nuclear warheads or even strategic nanoweapons warheads. This is obviously the wrong direction. We need to lower tensions and to prevent the use of WMD. Therefore, it is imperative that everyone agree to keep hypersonic glide missiles confined to conventional warheads. In my view, this is an urgent area for the United Nations to focus on.

2. Define autonomous smart nanobots (ASN) as equivalent to nuclear weapons and falling under the Treaty on the Non-Proliferation of Nuclear Weapons and the Limited Test Ban Treaty. My reasoning is that both are WMD and threaten human extinction. There is no doubt that ASN have the potential to cause destruction equivalent to nuclear weapons. Similar to nuclear weapons, the amount of destruction is programmable. In my view, the United Nations should work with its members to define autonomous smart nanobots as equivalent to nuclear

weapons and categorize them as falling under the existing nuclear weapons treaties.

3. Allow nations to build antiballistic missile systems, if they choose to do so. Today we count on the doctrine of MAD (mutually assured destruction) to avert nuclear war. To date, that has worked. However, as weapons grow more complex, the possibility of an unintentional missile launch grows greater. We need to ask ourselves: Do we want to risk destroying the planet in the advent of an unintentional missile launch? Even if some adversary, such as a rogue state, intentionally launched intercontinental ballistic missiles, do we want our only defense to be nuclear retaliation? Consider carefully. It is a no-win situation. In the words of President Ronald Reagan:

"For these reasons and because of the awesome destructive potential of nuclear weapons, we must seek another means of deterring war. It is both militarily and morally necessary. Certainly, there should be a better way to strengthen peace and stability, a way to move away from a future that relies so heavily on the prospect of rapid and massive nuclear retaliation and toward greater reliance on defensive systems which threaten no one."

After having surveyed the available literature, I believe Reagan is correct. As of this writing, both China and North Korea are improving their missile delivery systems. In the coming decade, I believe they will both have accurate intercontinental ballistic missiles (ICBMS). To my mind, all nations need to neutralize the threat that ICBMS pose. From a basic scientific viewpoint, it is not obvious that the technology for a completely effective antiballistic missile system exists. The technology may take a decade or more to mature and deploy. Even when it does exist, nanoweapons will be the new weapons of mass destruction, and there is evidence that suggests their delivery systems will not depend on ICBMS. I do not think we can stop nations from building antiballistic missile systems, and nanoweapons may negate their effectiveness as a deterrent. In effect, my suggestion is that we accept what is likely to be inevitable.

My expertise is in technology, not political policy or ethics. I am viewing this from the standpoint of a technologist and recommending strategies based on the similarity of nanoweapons to other weapons that threaten human extinction. My view is that we already have experience with biological and nuclear weapons that threaten our extinction and have effectively curtailed their use via multilateral treaties and conventions. Let's take advantage of our experience with existing WMD and use it to curtail the use of strategic nanoweapons.

To summarize, my suggestions are straightforward:

1. Classify weaponized self-replicating smart nanobots (SSN) as a life-form (artificial life) and categorize it as part of the Biological Weapons Convention.

2. Confine hypersonic glide missiles to deploy only conventional weapons.

3. Define autonomous smart nanobots as equivalent to nuclear weapons, since they are both weapons of mass destruction, and categorize them as falling under the Treaty on the Non-Proliferation of Nuclear Weapons and the Limited Test Ban Treaty.

4. Allow nations to build antiballistic missile systems if they choose to do so.

These are only suggestions. They are conceptually simple, but I believe the logic that underpins them is sound. I recognize that there is much work ahead, and the international treaties may take many different forms. In the final analysis, it will be a journey. Leaders who respect the fundamental right that all humans have to life will lead the way. I view this book as the first step on that journey. You have taken that step. You have taken the time to become informed. You now know that the next and most deadly generation of military weapons the world has ever encountered, nanoweapons, threatens human extinction.

Now this is not the end. It is not even the beginning of the end. But it is, perhaps, the end of the beginning.

—WINSTON CHURCHILL

APPENDIX 1

Institute for Soldier Nanotechnologies

The Institute for Soldier Nanotechnologies (ISN) website (http://isnweb
.mit.edu) lists five "strategic research areas," delineated as SRAS.
Under each SRA they list themes and projects. Because the list
demonstrates the depth and breadth of the Army's nanoweapon
research, this appendix reproduces it verbatim from the ISN
website.

SRA 1: Lightweight, Multifunctional Nanostructured Materials

Theme 1.1: Quantum Dots for Wide-Bandwidth Imaging
and Communications

Project 1.1.1: Hybrid Quantum Dot-Based Imagers and
Emitters with Broadly Tunable Spectral Characteristics

Theme 1.2: Nanoscale Carbon Forms for Situational
Awareness

Project 1.2.1: Graphene Devices for Next-Generation Night
Vision Systems

Theme 1.3: High-Functionality Nanostructured Surface
Capabilities

Project 1.3.3: Enabling Architectures and Technologies
for Next-Generation Fiber Devices

Theme 1.4: Environmental Obfuscation and Extended Reach
Situational Awareness

Project 1.4.1: Tailored Nano-particles for Obscurant
Applications

SRA 2: Soldier Medicine—Prevention, Diagnostics, and Far-Forward Care

Theme 2.1: Cellular Immune Response and Nano-engineered Drug Delivery

Project 2.1.1: Nanotechnology for Stimulating, Sampling, and Monitoring Immunity

Theme 2.2: Microfluidics of Dissolution and Mixing

Project 2.2.1: Rapid Reconstitution Packages of Lyophilized Medicines

Theme 2.3: Synthesis and Characterization of Therapeutic Materials

Project 2.3.1: Nano-structured Biomaterials for Treatment of Hemorrhagic Shock

Project 2.3.2: Multi-component Nanolayer Assemblies for Soldier Wound Healing

Project 2.3.3: Delivery of Brain Lipid Nanoparticles Using Microtech Devices for Treatment of Traumatic Brain Injury

Project 2.3.4: Complementary Wound-Healing Strategies Enabled by Synthetic Biology and Nanotechnology

SRA 3: Blast and Ballistic Threats: Materials Damage, Injury Mechanisms, and Lightweight Protection

Theme 3.2: Grain and Phase Boundary Manipulation in Nanocrystalline Metal Alloys

Project 3.2.1: Layered/Graded Nanocrystalline and Superelastic-Fiber Alloys for Lightweight Protection

Theme 3.3: Physical, Biological, and Physiological Mechanisms of Injury and Tissue Materials Modeling

Project 3.3.1: Blast-Induced TBI—Connections among the Physical, Biological, and Behavioral Dimensions

Project 3.3.2: Electromechanical Interactions in Blast-Induced Traumatic Brain Injury

Project 3.3.3: Molecular to Macroscale Exploration of Fundamental Properties of Gels

Theme 3.4: Multi-scale Descriptions of Complex Mechanical Phenomena

Project 3.4.1: Advanced Computational Tools for Multiscale Modeling and Simulation of Multi-Threat Protective Systems

Theme 3.5: Novel Bio-inspired and Carbon-nanostructured Materials

Project 3.5.2: Design and Synthesis of Carbon-based Chainmail Structures for Flexible, Ultra-lightweight Protection

SRA 4: Hazardous Substances Sensing

Theme 4.1: Coupled Optoelectronic and Biochemical Mechanisms in Specialized Nanostructures

Project 4.1.2: Resistivity-Based Microfluidic Biosensing

Project 4.1.4: Molecular Recognition Using Carbon Nanotube Adsorbed Polymer and Bio-Polymer Phases: Synthetic Nanotube Templated Antibodies

Theme 4.2: Quantum Dots for Chem/Biosensing

Project 4.2.1: Chem-Bio Analyte Sensing with Hybrid Quantum Dot Constructs

SRA 5: Nanosystems Integration: Flexible Capabilities in Complex Environments

Theme 5.1: Coupled Optoelectronic and Phononic Structure in Novel Fiber Materials

Project 5.1.1: Ferroelectric Acoustic Fibers

Theme 5.2: New Regimes of Light Matter Interactions in Complex Nanostructured Materials

Project 5.2.1: Multifunctional Integrated Fabrics

Project 5.2.2: Enabling Novel Lightwave Phenomena

Project 5.2.3: Spatial Awareness around Corners

Theme 5.3: Novel Thermal Radiation Phenomena in Photonic Crystal Materials Systems

Project 5.3.1: Novel Thermal Radiation Management Using Advanced Photonic Crystals

Nanoweapons Offensive Capability of Nations (NOCON)

Categories

To bring order to our discussion, this list has three categories:

1. Nanoweapons nations: Nations that are already deploying nanoweapons and have robust development programs.

2. Near follower nanoweapons nations: Nations that are developing or deploying nanoweapons but that are one or more generations behind nanoweapons nations

3. Nanoweapons capable nations: Nations that are capable of developing nanoweapons but have chosen not to do so.

Methodology Overview

The methodology to assign nations to a NOCON category requires a fourfold approach:

1. The Cientifica report, "Nanotechnology Funding: A Global Perspective," by Tim Harper, was presented at the National Nanotechnology Initiative Workshop in May 2012. It delineates the Nanotechnology Impact Factor (NIF) of key nations. The NIF measures how the impact of a nation's thrusts in nanotechnology is likely to affect its economy. The NIF list is the first critical piece of information.

2. To understand how a nation's thrusts in nanotechnology will affect its ability to develop nanoweapons, we examine the 2015 defense budgets of nations and consider it representative of the nation's military industrial complex. The defense budgets are the second critical pieces of information.

3. To develop a metric that expresses a nation's capability to develop and deploy offensive nanoweapons, which we will term the Nanoweapons Offensive Capability of Nations or NOCON score, we use the following equation:

NOCON score = (military budget) × (NIF expressed as a decimal)

This, in effect, says that a nation's military budget is representative of their military industrial complex and a nation's Nanotechnology Impact Factor (NIF, expressed as a decimal relative to the highest being 1.0) determines its ability to develop and deploy offensive nanoweapons, the NOCON score.

4. We will also examine national alliances that enable nanotechnology and nanoweapons sharing.

5. Combining the results from items 3 and 4, we will use judgment to develop the Nanoweapons Offensive Capability of Nations list.

Applying the Methodology

We will start by "following the money." Harper indicates that of the quarter trillion dollars spent on nanotechnology between 2001 and 2012, the countries that invested most are, in descending order:

1. United States

2. Russia

3. China

4. European Commission, the executive body of the European Union

5. Japan

6. India

In terms of total spending, adjusted for purchasing power parity (PPP), the United States, Russia, and China are investing about the same regarding nanotechnology investments. The European Commission is investing about a third as much as the top three, Japan about a quarter, and India about a tenth.

In addition to adjusting investments in terms of PPP, Harper also adjusts investments in terms of the Emtech Exploitation Index (EEI). The EEI is a measure regarding how effectively a country can "commercially" exploit its nanotechnology investments. It takes into account factors such as overall global competitiveness, quality of scientific institutions, and capacity for innovation. Here are the EEIS provided in the Cientifica report for various nations:

1. United States   5.00
2. Germany   4.93
3. Taiwan   4.90
4. Japan   4.88
5. South Korea   4.60
6. United Kingdom   4.55
7. China   4.30
8. France   4.23
9. India   3.95
10. Russia   3.57

Cientifica concludes that the United States has the best infrastructure to capitalize on its investments. The report places emphasis on experienced researchers and scientists, as well as commercial, industrial, and medical "know how" to bring new products to market. From this, they conclude that Russia and China would have to spend almost twice as much in real PPP terms to outpace the United States in nanotechnology commercialization.

Combining national investments with EEIS, Cientifica develops a Nanotechnology Impact Factor (NIF), which measures how the impact of a nation's thrusts in nanotechnology is likely to affect its economy. The Cientifica NIFS reported below are "normalized" such that the highest score equals 100 and is in the first column (#1). The second column (#2) is the score represented as a decimal:

Nation	No. 1	No. 2
1. United States	100	1.00
2. China	89	0.89
3. Russia	83	0.83
4. Germany	30	0.30
5. Japan	29	0.29
6. EU	27	0.27
7. South Korea	25	0.25
8. Taiwan	9	0.09
9. UK	6	0.06
10. India	5	0.05

This information agrees with and expands the information presented in chapter 4, which focused on only the top three nations. However, we need to understand how a nation's thrusts in nanotechnology will affect its ability to develop nanoweapons. There is no "standard" methodology to do this. Therefore, we will need to develop a reasonable approach to make this determination.

Let's start by examining the 2015 defense budgets of the nations on the list. We will use the 2016 edition of *The Military Balance* published by the International Institute for Strategic Studies (IISS), which is a British research firm that focuses on international affairs. I have chosen to use the IISS because it provides military expenditures for 2015 using average market exchange rates and is one of the top ten most influential think tanks in the world. Accessing other think tank estimates will yield different results, but for the most part the general rankings are close. In addition, we will look at the military budgets of only the top ten nations on Cientifica NIFS report. We will also only focus on nations within the European Union (EU) that have the largest budgets. All 2015 budgets are in billions of U.S. dollars:

1. United States   597.5
2. China   129.8
3. United Kingdom   56.2
4. Russia   51.6
5. India   47.9
6 France   46.7
7. Germany   36.6
8. South Korea   33.4
9. Japan   11.9
10. Taiwan   10.2

Using a nation's NIF and military budget, we can develop a metric that expresses a nation's capability to develop and deploy offensive nanoweapons, which we term the "NOCON score." Further, we can define a nation's NOCON score according to the equation:

NOCON score = (military budget) x (NIF expressed as a decimal)

This, in effect, says that a nation military budget and commercial, industrial, and medical infrastructure (NIF) determines its ability to develop and deploy offensive nanoweapons. Since the Cientifica report did not provide a NIF for France, we will equate it to the European Union's NIF. In general, this appears conservative:

- France's nanotechnology thrusts benefit from its membership in European Framework Programmes for Research and Technological Development

- France has the second highest military budget within the European Union

If we apply the NOCON score equation, we get the following:

1. United States   597.5
2. China   115.5
3. Russia   43.8
4. France   12.6

5. Germany	11.0
6. South Korea	8.4
7. United Kingdom	3.4
8. India	2.4
9. Japan	3.4
10. Taiwan	0.9

Results

The above list provides insight regarding how to categorize nations in NOCON. However, it cannot be "strictly by the numbers." Military alliances will also play a role. This is especially true of the United Kingdom, which has the largest military budget of all European Union nations and ranks fifth in the world overall, even ahead of Russia's military budget. This begs a question, Why don't they rank higher on the NOCON score? This is a direct result of their low NIF ranking. My interpretation of the Cientifica report suggests that they attribute the United Kingdom's low NIF results to three factors:

1. A lack of strategic focus and continuity

2. Environment, Health and Safety (EHS) concerns

3. Poor synergy between academic research and industry

While I respect Cientifica's analysis, I do not agree. Consider this. The United Kingdom is the United States' closest ally. In fact, military strategists characterize their relationship as "Special." The level of cooperation between them with regard to military planning, execution of military operations, nuclear weapons technology, and intelligence sharing is "unparalleled" among major powers. There is every reason to believe that this "special relationship" includes nanotechnology and nanoweapons sharing. In addition, the United Kingdom is a nuclear power and a NATO nation. It would be in the best interest of the United States to ensure the United Kingdom is a fully capable military ally, including its ability to deploy nanoweapons. However, this is not a one-sided relationship. The United Kingdom would be able to share any advances in nanotechnology it makes resulting from its membership to the European Union's nanotechnology development thrusts of European Framework Programmes for Research and Technological Development. From an economic viewpoint, including trade and commerce, and from a military viewpoint, the United Kingdom is in my view the "undeclared 51st state" of the United States. I do not make this statement lightly. The United Kingdom and the United States have been close allies for well over a hundred years, and their "special relationship" has been forged in numerous military and political conflicts such as World War I, World War II, the Korean War, the Cold War, the Falklands War, the Gulf Wars, and now the War on Terror.

Let's also discuss Saudi Arabia. They did not show up on any of Cientifica's lists. Based on that, it would be reasonable to believe they are not pursuing nanoweapons. However, the 2016 edition of *The Military Balance* published by IISS estimates their 2015 military budget at $81.8 billion, which is the third largest military budget in the world, behind only the United States and China. They also are a close ally of Pakistan, sharing commercial, cultural, religious, and political interests. Pakistan is a nuclear power. Currently, Saudi Arabia's focus is on remaining one of the world's largest oil exporters. Their foreign policy appears guided by maintaining relations with other oil-producing and major oil-consuming countries. However, the Middle East is a turbulent hotbed for religious conflict, as well as wars related to the region's oil resources and access to the Persian Gulf. For example, one major reason Iraq invaded the small nation of Kuwait in the first Gulf War was to gain direct access to the Persian Gulf. In a practical sense, Saudi Arabia has a large defense budget and aligns itself with Pakistan to deal with their perceived threat from the Islamic Republic of Iran and other hostilities in the region. However, that is not the whole story. The Middle East is the most militarized region in the world. Saudi Arabia's relationship with the United States draws criticism from radical Islamic groups throughout the region. Given Saudi Arabia's military budget, its precarious position in the Middle East, and its alliance with Pakistan, it would be reasonable to believe they could acquire nanoweapons. Today, most arms sales are to Middle East nations. I am not suggesting Saudi Arabia is developing nanoweapons. I am suggesting their oil wealth and defense budget would enable them to buy some of the best nanotechnologists in the world to develop offensive nanoweapons, or even buy offensive nanoweapons that become available in the military market, if they perceive it in their national interest.

Conclusions and Rationale

Given the above information, this is how I view the Nanoweapons Offensive Capability of Nations:

1. Nanoweapons Nations

• United States is the clear leader in developing and deploying offensive nanoweapons. Close alliances with NATO nations like the United Kingdom and France may prove synergistic in developing nanotechnology and nanoweapons.

• China is a near follower that understands the importance of nanoweapons, deploys them as they become available, and is ramping up its nanoweapons programs as rapidly as its economy allows. Alliances with Russia and North Korea may prove synergistic in developing nanotechnology and nanoweapons.

• Russia is a distant third that has gotten a relatively poor return on its investment in nanotechnology, mainly due to corruption and poor management. Its alliance with China may prove synergistic in developing nanotechnology and nanoweapons.

• United Kingdom: for reasons previously stated.

2. Near Follower Nanoweapons Nations

France is likely developing nanoweapons via the Centre National de la Recherché Scientifique and participation in the nanotechnology development thrusts of European Framework Programmes for Research and Technological Development. They are a NATO nation and a nuclear power. French nuclear weapons, like those of the United States, United Kingdom, and Russia, are on high alert, ready to use on short notice. Their nanoweapons would likely result from their commercial applications, for example, using nanoparticles for automotive catalytic converters, and their ever increasingly close alliance with the United States. Some military analysts suggest that France is replacing the United Kingdom as the United States' closest ally. I do not agree, but I do feel that France equals the United Kingdom in military capability. They also politically align with the United States regarding defeating Islamist militants and opposing Russian aggression. In fact, in 2013 French president François Hollande launched a military intervention in Mali to defeat Islamist militants.

• Germany is likely developing nanoweapons. As of 2015, there are about a thousand German companies engaged in the development, application, and distribution of nanotechnology products. Those companies account for about 70,000 jobs. In general, Germany's nanotechnologies thrusts are increasing. Historically, Germany is a signatory of the Treaty of Brussels, which prohibited them to possess nuclear, biological, or chemical weapons. The Brussels treaty expired in 2010, and the North Atlantic Treaty Organization (NATO) replaced it. Germany is a member of NATO. Germany is also a signatory of the Non-Proliferation Treaty and agrees not to develop or deploy strategic nuclear weapons. However, as a NATO nation under the "United States–NATO nuclear weapons sharing" provisions, Germany, along with Belgium,

Italy, the Netherlands, and Turkey, receives tactical nuclear weapons from the United States, which also controls their use. Given Germany's strong nanotechnology infrastructure and constraints regarding nuclear, biological, and chemical weapons, it is logical that they would seek to have a strong position in nanoweapons. I think it is likely, for example, that Germany is developing specific nanoweapons, such as mini-nukes. Germany is also a member of the European Union and participates in the nanotechnology development thrusts of European Framework Programmes for Research and Technological Development.

• South Korea is the world's fifteenth largest economy and has nearly 30,000 U.S. troops stationed within its borders. South Korea is an ally of the United States and an important trade partner. In addition, South Korea's military budget is almost as large as the entire GDP of North Korea, its hostile neighbor. It is likely that:

> The United States troops in South Korea have the latest generation of conventional weapons and tactical nuclear weapons to thwart a nuclear strike by North Korea.

> South Koreas is developing offensive nanoweapons, especially related to the threat North Korea poses.

3. Nanoweapons Capable Nations

Japan primarily focuses its nanotechnology on commercial, industrial, and medical applications. Based on the 1960 Treaty of Mutual Cooperation and Security between the United States and Japan, Japan is a U.S. protectorate and an important trading partner. Strategically it makes sense for Japan to focus on its continued economic growth and strengthening its relationship with the United States. Most military strategists suggest that Japan would like to upgrade from protectorate to ally.

India primarily focuses its nanotechnology on commercial, industrial, and medical applications. India is a nuclear power. It is actively pursuing new ballistic missile, cruise missile, and sea-based nuclear delivery systems. Given India's low military budget and modest nanotechnology capabilities, it is doubtful that there is a robust nanoweapons program.

Saudi Arabia: for reasons previously stated.

The Events Leading to the Chernobyl Disaster

1. At 1:06 a.m. (local time) April 25, 1986, a gradual reduction in the output of the power of Chernobyl's reactor #4 commenced until the power level reached 50 percent of its nominal 3,200 MW thermal level. This took a number of hours, and by the time the power level reached 50 percent, the day shift had begun. This first step proceeded without incident.

2. Suddenly, without warning, a regional power station went offline. This prompted the Kiev electrical grid controller to request postponing any further reduction of Chernobyl's output. The controller's concern was to have sufficient power for the peak evening demand. This request was unexpected, but did not appear to represent a threat to safety. Chernobyl complied.

3. Preparations for the test not affecting the reactor's power continued, including disabling the emergency core cooling system. The ECCS is a backup system intended to provide cooling water to the core in a loss-of-coolant accident. The disabling of the ECCS, in hindsight, represented a questionable judgment and the inherent lack of attention to safety.

4. After satisfying the evening peak demand, at 11:04 p.m. the Kiev grid controller gave permission to resume the Chernobyl reactor shutdown. At this point, the evening shift was getting ready to leave. The night shift would take over at midnight. The original plan required the test to be completed during the day shift. The night shift's job was only "maintenance" of the decay heat cooling systems in a shutdown plant. A large "red flag" should have been apparent and caused a halt of the

test. The night shift was ill prepared to carry out any elements of the plan, other than maintenance. Still, no one voiced concerns, and the test proceeded.

5. During the shift change, an inexperienced night shift engineer executed a rapid reduction in the power level from 50 percent to a little over 20 percent. By 12:05 a.m. on April 26, the reactor was at 700 MW test goal level.

6. Then the unexpected happened: the core power continued to decrease. The nuclear physics regarding this behavior is well known. In the parlance of nuclear physics, it is a process known as "reactor poisoning." (Please consult the glossary and annotations for more information. Understanding the exact nuclear physics of reactor poisoning is unnecessary. It is only important to know that the engineer should have anticipated reactor poisoning given the conditions surrounding the rate of power reduction.)

7. When the reactor power reached 500 MW, the engineer mistakenly inserted the control rods too far. We will never know the cause of this error, as the engineer and the night shift manager died shortly following the disaster.

8. This rendered the reactor in a near shutdown mode, with a power output of less than 30 MW, or about 5 percent of the output deemed safe for the test.

9. To counter the reactor poisoning, the control room personnel decided to disable the automatic system governing the control rods and manually extract the reactor control rods to their upper limits, but to no avail. They were unable to get the reactor above 200 MW, which allowed increased poisoning of the reactor core.

10. The reactor's low operational level and increased poisoning triggered alarms. Between 12:35 and 12:45 a.m., the control room personnel ignored emergency alarm signals concerning thermal-hydraulic parameters, apparently to maintain the reactor

power level. In a decision that defies reason, at 200 MW and with increased reactor poisoning, the control room continued the test.

11. Per the test plan, the control room activated extra water pumps at 1:05 a.m. The increased water flow through the reactor produced an increase in the inlet coolant temperature of the reactor core because the water did not have sufficient time to release its heat in the turbine and cooling towers. This reduced the cooling efficiency and safety margin.

12. The increased water flow exceeded the allowed limit. As a result, at 1:19 a.m. an alarm signaled low steam pressure in the steam separators. Since water absorbs neutrons better than steam, turning on the additional pumps decreased the reactor's power even further. The crew responded by turning off two pumps in an effort to increase steam pressure.

13. The result was an extremely unstable reactor configuration. Out of 211 control rods, the control room removed all but 18. Safety requires 28 rods in the reactor at all times. All automatic safeties were disabled. The control crew did not "manually" reinsert all the control rods to shut down the reactor. Their lack of action remains a mystery.

14. In the absurdity of this situation, they continued the experiment. At 1:23:04 a.m., with only four of the eight Main Circulating Pumps (MCP) active (six are normally active), the control room turned off the steam to the turbines, which allowed the turbine generator to run down. The diesel generators were to have completely picked up the MCPS' power needs by 1:23:43 a.m. In the interim, 31 seconds, the turbine generator, now coasting down, was supposed to power the MCPS.

15. When they turned off the turbine generator, it stopped powering the pumps. This caused the water flow rate to decrease, leading to the formation of more steam voids (bubbles) in the core. The formation of steam voids reduced the amount of the liquid water to absorb neutrons. This increased the reac-

tor's power output. The situation was now out of control. As the reactor's power increased, so did the steam voids. As the steam voids increased, the reactor's power increased.

16. At 1:23:40 a.m., the records indicate that someone manually pressed the EPS-5 button of the reactor emergency protection system, which automatically inserted all control rods.

17. The design of the control rods included a graphite tip, which displaced neutron-absorbing water with graphite before introducing neutron-absorbing boron material. Initially, this increased the reaction rate.

18. A few seconds after the graphite rod tips entered the fuel pile, a massive power spike occurred. This caused the core to overheat, fracturing some of the fuel rods. The fractured fuel rods blocked the control rod columns at a point that placed the graphite tips in the middle of the core. The reactor output shot above 530 MW in three seconds.

19. Instruments did not record the rest of what happen. We only know that the last reading on the control panel was 33,000 MW, ten times the normal output. Here is one simulation of what might have happened:

a. The power spike caused an increase in fuel temperature.

b. The increase in fuel temperature caused massive steam buildup.

c. The massive steam buildup caused a rapid increase in steam pressure.

d. The rapid increase in steam pressure caused the fuel cladding to fail.

e. This allowed the fuel and water to mix.

f. This was followed by three steam explosions, two internal to the reactor and the final one external.

g. The final explosion tore off the 1,000-ton upper plate, which is fastened to the entire reactor, and shot it through the roof of the building.

21. Witnesses report huge fireball eruptions 1,000 feet into the air that continued for two days. The actual fire lasted seven days, as the reactor melted down.

NOTES

Introduction

xi **The events that most people consider**: A. Sandberg and N. Bostrom, "Global Catastrophic Risks Survey," Technical Report #2008-1 (New York: Oxford University Press, 2008).

1. What You Don't Know Can Kill You

8 **Fact 1: The National Nanotechnology Initiative**: "NNI Vision, Goals, and Objectives," Nano.gov, retrieved April 13, 2016, http://www.nano.gov/about-nni/what/vision-goals.

10 **Given these facts, it is not**: David M. Berube, "Public Perception of Nano: A Summary of Findings," *NanoHype: Nanotechnology Implications and Interactions*, retrieved April 13, 2016, http://nanohype.blogspot.com/2009/10/public-perception-of-nano-summary-of.html#uds-search-results.

11 **In 2001 Nobel laureate Richard Smalley**: Richard E. Smalley, "Of Chemistry, Love, and Nanobots," *Scientific American* 285, no. 3 (September 2001): 76–77.

12 **In 2007 the Russian military successfully**: Adrian Blomfield, "Russian Army 'Tests the Father of All Bombs,'" *Telegraph*, September 12, 2007, retrieved April 13, 2016, http://www.telegraph.co.uk/news/worldnews/1562936/Russian-army-tests-the-father-of-all-bombs.html.

2. Playing LEGOS with Atoms

25 **Nanolithography is also a technique used**: M. Despont, J. Brugger, U. Drechsler, U. Dürig, W. Häberle, M. Lutwyche, H. Rothuizen, R. Stutz, and R. Widmer, "VLSI-NEMS chip for parallel AFM data storage," *Sensors and Actuators* 80, no. 2 (2000): 100–107.

26 **The red color emits from the**: J. Winter, "Gold Nanoparticle Biosensors," May 23, 2007, https://leelab.engineering.osu.edu/sites/nsec.osu.edu/files/uploads/WinterGoldNanoparticles.pdf.

3. I Come in Peace

29 **Defining a nanotechnology product is challenging**: E. O'Rourke and M. Morrison, "Challenges for Governments in Evaluating Return on Investment from Nanotechnology and Its Broader Economic Impact," paper from *International Symposium on*

Assessing the Economic Impact of Nanotechnology, 2012, Nano.gov, http://www.nano
.gov/sites/default/files/dsti_stp_nano201212.pdf.

29   The 2012 International Symposium on Assessing: *International Symposium on
Assessing the Economic Impact of Nanotechnology*, 2012, Nano.gov, http://www.nano
.gov/node/785.

31   In fact, many consider nanotechnology: M. Knell, "Nanotechnology and the Challenges of Equity, Equality, and Development," *Nanotechnology and the Sixth Technological Revolution: Yearbook of Nanotechnology in Society* 2 (September 30, 2010): 127–43.

31   Zettl's machines, like the nanoradio: K. Bourzac, "Nanoradio Tunes In to Atoms,"
MIT Technology Review, July 21, 2008, https://www.technologyreview.com/s/410487
/nanoradio-tunes-in-to-atoms.

32   Some estimates place the worldwide market: M. Roco, "Nanotechnology Research
Directions for Societal Needs in 2020," March 28, 2012, *International Symposium
on Assessing the Economic Impact of Nanotechnology*, http://www.nano.gov/node/797.

32   An estimate of the worldwide economy: A. Laudicina and E. Peterson, "Global
Economic Outlook, 2015–2020: Beyond the New Mediocre," January 2015, A. T.
Kearney Global Business Policy Council, https://www.atkearney.com/documents
/10192/5498252/Global+Economic+Outlook+2015-2020—Beyond+the+New+Mediocre
.pdf/5c5c8945-00cc-4a4f-a04f-adef094e90b8.

32   Given the stakes, economically and militarily: M. Roco, "Nanotechnology Research
Directions for Societal Needs in 2020," March 28, 2012.

33   The CPI was updated in 2013: M. E. Vance, T. Kuiken, E. P. Vejerano, S. P. McGinnis, M. F. Hochella, D. Rejeski, and M. S. Hull, "Nanotechnology in the Real World:
Redeveloping the Nanomaterial Consumer Products Inventory," *Beilstein Journal of
Nanotechnology*, 2015, 1769–80, http://www.beilstein-journals.org/bjnano/single
/articleFullText.htm?publicId=2190-4286-6-181.

34   Intel states on their website: Intel Corporation, retrieved April 16, 2016, http://www
.intel.com/content/www/us/en/processors/core/core-m-processors.html.

35   For example, did you know: Terry Paige, "The Rise of Nanotechnology," *LifeFact PIA*,
May 11, 2013, http://lifefactopia.com/technology/Nanotechnology.

35   We do know that the National: "Nano Exposed: A Citizen's Guide to Nanotechnology," *Organic Consumers*, December 2010, https://www.organicconsumers.org
/sites/default/files/nano-exposed_final_41541.pdf.

36   The 2016 NNI budget proposes: "Supplement to the President's Budget for Fis-
cal Year 2016," The National Nanotechnology Initiative, White House, https://
www.whitehouse.gov/sites/default/files/microsites/ostp/nni_fy16_budget_supple
ment.pdf.

36   The use of concrete-like materials: N. Gromicko and K. Shepard, "The History of
Concrete" *InterNACHI*, http://www.nachi.org/history-of-concrete.htm#ixzz31v47zuu.

36   The ancient Romans used concrete extensively: N. Gromicko and K. Shepard, "The
History of Concrete."

36   There is wide agreement that concrete: Florence Sancheza and Konstantin
Sobolevb, "Nanotechnology in Concrete," May 15, 2010, https://www.researchgate
.net/publication/222873024_Nanotechnology_in_Concrete_-_A_Review.

36 **Nanotechnology is finding wide application**: S. Mann, "Nanotechnology and Construction," Nanoforum.org, European Nanotechnology Gateway, October 31, 2006, http://www.nanowerk.com/nanotechnology/reports/reportpdf/report62.pdf.

36 **Steel has been used**: S. Mann, "Nanotechnology and Construction."

36 **In fact, according to a forecast**: Wesley Cook, "Bridge Failure Rates, Consequences, and Predictive Trends," *All Graduate Theses and Dissertations*, 2163, 2014, http://digitalcommons.usu.edu/etd/2163.

37 **Although this collapse was due**: S. Jones, "Friday Marks 7 Years since I-35w Bridge Collapse," Minneapolis (WCCO), August 1, 2014, http://minnesota.cbslocal .com/2014/08/01/friday-marks-7-years-since-i-35w-bridge-collapse; Charles C. Roberts Jr., "Minneapolis Bridge Collapse," http://www.croberts.com/minneapolis-bridge -collapse.htm.

37 **According to structural engineer Surinder Mann**: S. Mann, "Nanotechnology and Construction," Nanoforum.org, European Nanotechnology Gateway, October 31, 2006, http://www.nanowerk.com/nanotechnology/reports/reportpdf/report62.pdf.

37 **In general, there are two ways**: "Nano-engineered Steels for Structural Applications," *Nanowerk Spotlight*, May 10, 2010, http://www.nanowerk.com/spotlight /spotid=16203.php.

38 **In 2015 MIT *Technology Review***: K. Bullis, "Nano-Manufacturing Makes Steel 10 Times Stronger," *MIT Technology Review*, February 16, 2015, https://www.technology review.com/s/534796/nano-manufacturing-makes-steel-10-times-stronger.

38 **According to NNI, examples include**: "Manufacturing at the Nanoscale," Nano.gov. retrieved April 15, 2016, http://www.nano.gov/nanotech-101/what/manufacturing.

40 **According to NNI, "NNI agencies**: "Manufacturing at the Nanoscale," Nano.gov.

40 **It devotes an entire section**: Nano.gov, retrieved April 15, 2016, http://www.nano.gov/you /nanotechnology-benefits.

40 **According to the editor in chief**: M. Patil, D. S. Mehta, and S. Guvva, "Future Impact of Nanotechnology on Medicine and Dentistry," *J Indian Soc Periodontol* 12, no. 2 (May–August 2008): 34–40.

41 **Experts at the European Medicines Agency**: "Reflection Paper on Nanotechnology-Based Medicinal Products for Human Use," *European Medicines Agency*, June 29, 2006, http://www.ema.europa.eu/docs/en_gb/document_library/Regulatory_and _procedural_guideline/2010/01/wc5 000   69728.pdf.

41 **Researchers Edward Kai-Hua Chow**: E. Kai-Hua Chow and D. Ho, "Cancer Nanomedicine: From Drug Delivery to Imaging," *Science Translational Medicine* 5, no. 216 (December 18, 2013).

41 **According to the Kidlington Centre**: "Nano-Medicine Market Size Is Expected to Be Worth $130.9 Billion by 2016," Kidlington Centre, Kidlington, UK, March 17, 2015, *PRNewswire*, http://www.prnewswire.com/news-releases/nano-medicine -market-size-is-expected-to-be-worth-1309-billion-by-2016-296544211.html.

42 **According to Robert A. Freitas**: Robert A. Freitas Jr., "The Ideal Gene Delivery Vector: Chromallocytes, Cell Repair Nanorobots for Chromosome Replacement Therapy," *Journal of Evolution and Technology* 16, no. 1 (June 2007): 1–97, http:// jetpress.org/v16/freitas.pdf.

46 **In 2009 the DoD published**: Department of Defense, "Defense Nanotechnology Research and Development Program," Nano.gov, 2009, retrieved April 15, 2016, http://www.nano.gov/node/621.

47 **Radiation-hardened integrated circuits processing**: "Rad Hard Microelectronics," Honeywell, retrieved April 15, 2016, http://www51.honeywell.com/aero/common /documents/myaerospacecatalog-documents/Space/Rad_hard_Microelectronics _Products_and_Services.pdf.

47 **Commercial integrated circuits can withstand**: *The International System of Units*, 2008, http://physics.nist.gov/Pubs/SP330/sp330.pdf.

48 **The realities of modern warfare**: S. Sagadevan and M. Periasamy, "Recent Trends in Nanobiosensors and Their Applications," *Rev. Adv. Mater. Sci.* 36 (2014): 62–69, http://www.ipme.ru/e-journals/rams/no_13614/06_13614_suresh.pdf.

48 **The entire sensor package:** Brittany Sause, "Nanosensors in Space," *MIT Technology Review*, 2007, https://www.technologyreview.com/s/408190/nanosensors-in-space.

49 **In 2014 the University of Massachusetts**: "Eco-Friendly 'Pre-Fab' Self-Assembling Nanoparticles Could Revolutionize Nano Manufacturing," *Kurzweil Accelerating Intelligence News*, August 14, 2014, retrieved April 15, 2016, http://www.kurzweilai.net /eco-friendly-pre-fab-self-assembling-nanoparticles-could-revolutionize-nano-man ufacturing; Timothy S. Gehan et al., "Multi-Scale Active Layer Morphologies for Organic Photovoltaics through Self-Assembly of Nanospheres," *Nano Letters*, 2014 (DOI: 10.1021/nl502209s).

52 **In 2015 the Office of Naval Research**: Naval S&T Strategy, 2015, retrieved April 15, 2016, http://www.navy.mil/strategic/2015-Naval-Strategy-final-web.pdf.

53 **However, in 2014, X. Zhao**: X. Zhao, W. Fan, J. Duan, B. Hou, and J. Pak, "Studies on Nano-Additive for the Substitution of Hazardous Chemical Substances in Antifouling Coatings for the Protection of Ship Hulls," *Pharm Sci.*, July 2014; 27 (4 Suppl): 1117–22., PMID: 25016277.

54 **According to *Nanowerk*:** "U.S. Navy Eyes Nanotechnology for Ultimate Power Control System," *Nanowerk News*, July 21, 2015, retrieved April 15, 2016, http://www .nanowerk.com/nanotechnology-news/newsid=40827.php.

55 **In 2003 the U.S. Army**: The Institute for Soldier Nanotechnologies, retrieved April 15, 2016, http://isnweb.mit.edu.

56 **In 2012 scientists at MIT:** J. Lee, D. Veysset, J. P. Singer, M. Retsch, G. Saini, T. Pezeril, K. Nelson, and E. L. Thomas, "High Strain Rate Deformation of Layered Nanocomposites," *Nature Communications* 3, Article number: 1164 doi:10.1038/ncomms2166, November 6, 2012,

57 **In fact, one-third of medical evacuations**: Hal Bernton, "Weight of War: Gear That Protects Troops Also Injures Them," *Seattle Times*, February 13, 2011, http:// www.seattletimes.com/nation-world/weight-of-war-gear-that-protects-troops-also -injures-them.

57 **Scientists at the U.S. Department**: "Nano-Thin Invisibility Cloak Makes 3D Objects Disappear," *Nanowerk News*, September 18, 2015, retrieved April 16, 2016, http:// www.nanowerk.com/nanotechnology-news/newsid=41348.php.

58 **In fact, CNN reported**: D. Melvin, "No More Dodging a Bullet, As U.S. Develops Self-Guided Ammunition," CNN, April 29, 2015, http://www.cnn.com/2015/04/29 /us/us-military-self-guided-bullet.

58 **The nano coating is especially interesting**: K. Bullis, "Nano-Manufacturing Makes Steel 10 Times Stronger," MIT Technology Review, February 16, 2015, https://www technologyreview.com/s/534796/nano-manufacturing-makes-steel-10-times -stronger.

59 **Researchers at Sandia Laboratories**: J. Gartner, "Military Reloads with Nano-tech," MIT Technology Review, January 21, 2005, https://www.technologyreview .com/s/403624/military-reloads-with-nanotech.

59 **In 2009 Northrop Grumman announced**: P. Pae, "Northrop Advance Brings Era of the Laser Gun Closer," Los Angeles Times, March 19, 2009, http://articles.latimes .com/2009/mar/19/business/fi-laser19.

61 **Currently about a third of all**: C. Coren, "Obama Administration to Increase Drone Flights 50 Percent," Newsmax, August 17, 2015, http://www.newsmax.com /Newsfront/drones-50-percent-airstrikes-Air-Force/2015/08/17/id/670454.

61 **Some estimate that drones have killed**: J. Serle, "Monthly Updates on the Covert War Almost 2,500 Now Killed by Covert U.S. Drone Strikes since Obama Inaugu-ration Six Years Ago: The Bureau's Report for January 2015," Bureau of Investigative Journalism, February 2, 2015, https://www.thebureauinvestigates.com/2015/02/02 /almost-2500-killed-covert-us-drone-strikes-obama-inauguration.

61 **Artificially intelligent (AI) machines will equal**: L. Muehlhauser, "When Will AI Be Created?" Machine Intelligence Research Institute, May 15, 2013, https:// intelligence.org/2013/05/15/when-will-ai-be-created.

62 **As the Wall Street Journal reported**: J. Hook, "Americans Want to Pull Back from World Stage, Poll Finds," Wall Street Journal, April 30, 2014, http://www.wsj.com /articles/sb10001424052702304163604579532050055966782.

62 **Future drones**: David Axe, "From Bug Drones to Disease Assassins, Super Weap-ons Rule U.S. War Game," Wired, August 24, 1012, http://www.wired.com/2012/08 /future-warfare.

62 **Although the DoD is tight-lipped**: "Speed Is the New Stealth," Economist, June 1, 2013, retrieved April 16, 2016, http://www.economist.com/news/technology -quarterly/21578522-hypersonic-weapons-building-vehicles-fly-five-times-speed -sound.

63 **I think it is also appropriate**: John Gartner, "Military Reloads with Nanotech," MIT Technology Review, January 21, 2005, https://www.technologyreview.com/s/403624 /military-reloads-with-nanotech.

65 **According to the NNI website**: Frequently Asked Questions, "How Does This Spend-ing Compare to Other Countries?" Nano.gov, http://www.nano.gov/nanotech-101 /nanotechnology-facts.

66 **However, from the 2012 NNI workshop**: "NNI Workshop Agendas and Presen-tations," Nano.gov, file downloads, "Tim Harper, Cientifica Ltd. Nanotechnol-ogy Funding: A Global Perspective. | 2.3 MB," http://www.nano.gov/sites/default /files/pub_resource/global_funding_rsl_harper.pdf.

67 **These are the number**: "Nanotechnology Patents in USPTO," *StatNano*, retrieved April 15, 2016, http://statnano.com/report/0103.

67 **In 2007 Russian leaders established**: *Rusnano Corporation*, retrieved April 15, 2016, http://en.rusnano.com/about.

68 **As a result, Chubais spoke**: "Russian State Tech Fund Starts Again after Missteps," Reuters, June 13, 2013, http://www.reuters.com/article/russia-rusnano-shakeup -idusl5n0ep2i420130613.

68 **In 2015 Putin placed Leonid Melamed**: "Бывшего главу «Роснано» обвинили в растрате," tvrain.ru, retrieved April 15, 2016, https://tvrain.ru/articles/byvshego _glavu_rosnano_obvinili_v_rastrate-390379.

69 **On September 3, 2015, China celebrated**: B. Gertz, "The China Challenge: The Weapons the PLA Didn't Show," *Asia Times*, September 8, 2015, http://atimes.com/2015/09/the -china-challenge-the-weapons-the-pla-didnt-show.

70 **According to the *Washington Free Beacon***: B. Gertz, "China Tests Anti-Satellite Missile," *Washington Free Beacon*, November 9, 2015, http://freebeacon.com /national-security/china-tests-anti-satellite-missile.

71 **History suggests China will find**: U.S. Naval Institute Staff, "China's Military Built with Cloned Weapons," U.S. Naval Institute, October 27, 2015, http://news .usni.org/2015/10/27/chinas-military-built-with-cloned-weapons.

72 **For example, Russia sold China its**: F. Gady, "China to Receive Russia's S-400 Missile Defense Systems in 12–18 Months," *Diplomat*, November 17, 2015, http:// thediplomat.com/2015/11/china-to-receive-russias-s-400-missile-defense-systems -in-12-18-months.

72 **Russian leadership wants that technology**: F. Gady, "Russia's Secret New Weapon: Should the West Be Afraid?" *Diplomat*, July 1, 2015, http://thediplomat.com/2015/07 /russias-secret-new-weapon-should-the-west-be-afraid.

72 **In 2014, *DefenseReview.com* published**: D. Crane, "Russian Nano-Armor Coming in 2015 for Future Soldier 'Warrior Suit,' and Russian Spetsnaz (Military Special Forces) Already Running Improved 6B43 Composite Hard Armor Plates, New Plate Carriers and Combat Helmets, AK Rifle/Carbines, GM-94 Grenade Launchers and Other Tactical Gear in Crimea, Ukraine," April 23, 2014, http://www.defensereview .com/russian-nano-armor-coming-in-2015-and-russian-spetsnaz-military-special -forces-already-running-improved-6b43-composite-hard-armor-plates-new-plate -carriers-ak-riflecarbines-gm-94-grenade-launch.

72 **In a speech to the Federal Assembly**: "'No One Will Prevail over Russia Militarily': Putin Eyes $700bn to Advance Army," rt.com, December 13, 2013, retrieved April 15, 2016, https://www.rt.com/news/putin-address-military-russia-125.

73 **During the mid-2000s, both Russia**: S. M. Hersh, *The Samson Option* (New York: Random House, 1991), 220.

73 **High-ranking GRU defector**: Stanislav Lunev, *Through the Eyes of the Enemy* (Washington DC: Regnery, 1998); N. Horrock, "FBI Focusing on Portable Nuke Threat," UPI, December 20, 2001, http://www.upi.com/Top_News/2001/12/21/fbi -focusing-on-portable-nuke-threat/90071008968550.

73   **According to a 2005 MIT review**: J. Gartner, "Military Reloads with Nanotech," MIT *Technology Review*, January 21, 2005, https://www.technologyreview.com/s/403624 /military-reloads-with-nanotech.

5. The Rise of the Nanobots

79   **The history of military robots dates**: Jon Turi, "Tesla's Toy Boat: A Drone before Its Time," *engadget*, January 19, 2014, http://www.engadget.com/2014/01/19/nikola -teslas-remote-control-boat.

79   **In 1866 British engineer Robert Whitehead**: Patrick McSherry, "Whitehead Torpedo," *Spanish American War Centennial*, http://www.spanamwar.com/torpedo.htm.

80   **During the 1930s and into the**: Alexander Lychagin, "What Is Teletank?" *Odint Soviet News*, October 9, 2004, http://translate.google.com/translate?hl=en&langpair =ru%7cen&u=http://www.odintsovo.info/news/?id=1683.

81   **The Soviet Union also deployed**: Alexey Isaev, "1942, Battle of Kharkov," interview for *Echo of Moscow*, http://echo.msk.ru/programs/victory/50054; and "A Short History VNIIRT," essays, http://pvo.guns.ru/book/vniirt/index.htm#_Toc122708803.

81   **The German Wehrmacht deployed**: Goliath Demolition Tank on YouTube, https:// www.youtube.com/watch?v=zhK8LoPgPdA.

81   **However, Nazi Germany did deploy robotic**: N. Trueman, "The V Weapons," History Learning Site, April 2015 and December 2015, http://www.historylearning site.co.uk/world-war-two/world-war-two-in-western-europe/the-v-revenge-weapons /the-v-weapons.

81   **The United States began its military**: P. W. Singer, "Drones Don't Die—A History of Military Robotics," *HistoryNet*, May 5, 2011, http://www.historynet.com/drones -dont-die-a-history-of-military-robotics.htm.

82   **In 1944 the U.S. Army Air**: John F. Keane and Stephen S. Carr, "A Brief History of Early Unmanned Aircraft," *Johns Hopkins APL Technical Digest* 32, no. 3 (2013), https://www.law.upenn.edu/live/files/3887-keane-and-carr-a-brief-history-of -early-unmanned.

83   **More successful was the Army-developed**: Peter Warren Singer, "Drones Don't Die—A History of Military Robotics," *HistoryNet*, May 5, 2011, http://www.historynet.com /drones-dont-die-a-history-of-military-robotics.htm.

83   **The first, in 1962, was to**: Peter Warren Singer, *Wired for War* (New York: Penguin, 2009), 54.

83   **The second, in 1979, was to**: "Lockheed MQM-105 Aquila," *Directory of U.S. Military Rockets and Missiles*, 2002, http://www.designation-systems.net/dusrm /m-105.html.

83   **Aerial drones did not play**: Chris Cole, "Rise of the Reapers: A Brief History of Drones," *Drone Wars UK*, June 10, 2014, http://dronewars.net/2014/10/06/rise-of-the -reapers-a-brief-history-of-drones.

83   **In 1999 the United States successfully**: R. Dixon, "UAV Employment in Kosovo: Lessons for the Operational Commander," Naval War College, February 8, 2000, www .dtic.mil/cgi-bin/Gettrdoc?ad=ada378573.

84   **Since 2003 the U.S. military**: K. Kawamura, D. M. Wilkes, and J. A. Adams, "Center for Intelligent Systems at Vanderbilt University: An Overview," *IEEE Systems, Man and Cybernetics Newsletter* 1, no. 3 (2003).

84   **In 2005 the U.S. Army**: Lewis Page, "Flying-Rifle Robocopter: Hovering Sniper Backup for U.S. Troops," *Register*, April 21, 2009, http://www.theregister.co.uk/2009/04/21/arss_hover_sniper.

84   **In a 2014 media release**: "The Future Is Now: Navy's Autonomous Swarmboats Can Overwhelm Adversaries," Office of Naval Research, 2014, retrieved April 16, 2016, http://www.onr.navy.mil/Media-Center/Press-Releases/2014/autonomous-swarm-boat-unmanned-caracas.aspx.

84   *Defense One* **reported**: Patrick Tucker, "The Pentagon Is Nervous about Russian and Chinese Killer Robots," *Defense One*, December 14, 2015, http://www.defenseone.com/threats/2015/12/pentagon-nervous-about-russian-and-chinese-killer-robots/124465.

84   **According to U.S. Deputy Secretary**: Tucker, "The Pentagon Is Nervous about Russian and Chinese Killer Robots."

84   **For example, the U.S. military is**: John W. Whitehead, "Roaches, Mosquitoes, and Birds: The Coming Micro-Drone Revolution," *HuffPost Tech*, June 17, 2013, http://www.huffingtonpost.com/john-w-whitehead/micro-drones_b_3084965.html.

84   **On February 2, 2016, DARPA announced**: "FLA Program Takes Flight," *Outreach@DARPA.MIL*, February 12, 2016, http://www.darpa.mil/news-events/2016-02-12.

85   **For example, on December 16, 2014**: Patrick Tucker, "The Military Wants Smarter Insect Spy Drones," *Defense One*, December 23, 2014, http://www.defenseone.com/technology/2014/12/military-wants-smarter-insect-spy-drones/101970.

85   **On May 15, 2015, Pfizer revealed**: "Pfizer Partnering with Ido Bachelet on DNA Nanorobots," *Next Big Future*, May 15, 2015, http://nextbigfuture.com/2015/05/pfizer-partnering-with-ido-bachelet-on.html.

85   *Next Big Future* **reports, "Bachelet has**: "Pfizer Partnering with Ido Bachelet on DNA Nanorobots."

86   **According to** 3tags.org, **Bachelet forms**: "DNA Nanobots Will Target Cancer Cells in the First Human Trial Using a Terminally Ill Patient," *3tags*, 2016, http://3tags.org/article/dna-nanobots-will-target-cancer-cells-in-the-first-human-trial-using-a-terminally-ill-patient.

86   **Jack Andraka, science prodigy and winner**: Joshua Ostroff, "Jack Andraka Invented a Cancer Breakthrough. Now He's Building Nanobots. He's 18," *Huffington Post*, July 15, 2015, http://www.huffingtonpost.ca/2015/07/09/jack-andraka-cancer-nanobots-treatment_n_7746760.html.

86   **Other medical researchers are taking**: Renier J. Brentjens et al., "CD19-Targeted T-Cells Rapidly Induce Molecular Remissions in Adults with Chemotherapy-Refractory Acute Lymphoblastic Leukemia," *Science Translational Medicine* 5, no. 177 (March 20, 2013), http://stm.sciencemag.org/content/5/177/177ra38.

87   **The technology to determine DNA ancestry**: *Ancestry.com*, retrieved February 17, 2016, http://dna.ancestry.com/?hl=Explore+your+heritage+with+dna&s_kwcid=dna+testing+for+heritage&gclid=Cj0keqiArou2brdcoN_c6ndi3ombeiqaneix5hOezsgzY

_edMrLuu1jlrxsgkwrohzythytSojl_Qt8aApyo8p8haq&o_xid=55307&o_lid=55307&o
_sch=Paid+Search+%e2%80%93+NonBrand.

6. The "Swarm"

91 **We see evidence of swarming's effectiveness:** Roland Bouffanais, *Design and Control of Swarm Dynamics* (New York: Springer, 2015).

92 **If you have any doubt:** "Pandemic Flu History," Flu.gov, http://www.flu.gov/pandemic/history.

92 **The bee's ability to sense:** Tracy V. Wilson, "Dinner and Dancing: Bee Navigation," *How Stuff Works*, May 30, 2007, http://animals.howstuffworks.com/insects/bee5.htm.

93 **The U.S. armed forces, regardless:** "Learn the 11 Military General Orders," Military.com, http://www.military.com/join-armed-forces/military-general-orders.html.

94 **Perhaps such nanobots would have:** "DNA," *The Encyclopedia of Earth*, http://www.eoearth.org/view/article/158858.

94 **In each case, it involved:** "Swarms of DNA Nanorobots Execute Complex Tasks in Living Animal," Foresight Institute, http://www.foresight.org/nanodot/?p=6410.

95 **The Faculty of Life Sciences:** "Universal Computing by DNA Origami Robots in a Living Animal," *Nature Nanotechnology*, April 6, 2014, http://www.nature.com/nnano/journal/v9/n5/full/nnano.2014.58.html#affil-auth.

97 **Some infectious flu viruses can live:** "Preventing Seasonal Flu Illness," Centers for Disease Control and Prevention, retrieved April 16, 2016, http://www.cdc.gov/flu/about/qa/preventing.htm.

7. The "Smart" Nanoweapons

102 **No computer currently has:** Louis A. Del Monte "When Will a Computer Equal a Human Brain?" *Science Questions and Answers*, June 5, 2014, http://www.louisdelmonte.com/when-will-a-computer-equal-a-human-brain.

103 **In 1950 mathematician and computer scientist:** Alan Turing, "Computing Machinery and Intelligence," *Mind* 59, no. 236 (October 1950): 433–60.

103 **The most notable is the Eugene:** Lance Ulanoff, "The Life and Times of 'Eugene Goostman,' Who Passed the Turing Test," *Mashable*, June 12, 2014, http://mashable.com/2014/06/12/eugene-goostman-turing-test/#gHxdw0xr2iqQ.

104 **Most AI researchers predict this will:** Ben Rossington, "Robots 'Smarter than Humans within 15 Years,' Predicts Google's Artificial Intelligence Chief," *Mirror*, February 23, 2014, http://www.mirror.co.uk/news/technology-science/technology/ray-kurzweil-robots-smarter-humans-3178027; and Nadia Khomami, "2029: The Year When Robots Will Have the Power to Outsmart Their Makers," *Guardian*, February 22, 2014, http://www.theguardian.com/technology/2014/feb/22/computers-cleverer-than-humans-15-years.

104 **But what is Moore's law?:** Gordon Moore, "Progress in Digital Integrated Electronics," 1975 IEEE *Text Speech*, retrieved April 16, 2016, http://www.eng.auburn.edu/~agrawvd/course/e7770_Spr07/read/Gordon_Moore_1975_Speech.pdf.

105 **There is a lot of negative**: "Fully Autonomous Weapons," *Reaching Critical Will*, retrieved February 25, 2016, http://www.reachingcriticalwill.org/resources/fact-sheets /critical-issues/7972-fully-autonomous-weapons.

105 **Does a computer with intelligence equivalent**: "Do you think artificial intelligence can ever be equal to humans in judgement and emotions?" *Debate.org*, retrieved February 25, 2016, http://www.debate.org/opinions/do-you-think-artificial-intelligence -can-ever-be-equal-to-humans-in-judgement-and-emotions.

106 **For example, according to *Defense One***: Patrick Tucker, "The Pentagon Is Nervous about Russian and Chinese Killer Robots," *Defense One*, December 14, 2015, http:// www.defenseone.com/threats/2015/12/pentagon-nervous-about-russian-and-chinese -killer-robots/124465.

106 **In general, the U.S. military complies**: "United States of America, Practice Relating to Rule 139. Respect for International Humanitarian Law," *International Committee of the Red Cross*, retrieved April 16, 2016, https://www.icrc.org/customary-ihl /eng/docs/v2_cou_us_rule139.

110 **By 2045, most researchers and futurists**: Louis A. Del Monte, *The Artificial Intelligence Revolution* (North Charleston SC: Createspace, April 2014), 128.

8. The Genie Is Loose

111 **However, the U.S. Energy Department**: "Manhattan District History," U.S. Department of Energy, retrieved February 28, 2016, https://www.osti.gov/opennet/manhattan _district.jsp.

112 **There were 130,000 people working**: Barton J. Bernstein, "The Uneasy Alliance: Roosevelt, Churchill, and the Atomic Bomb, 1940–1945," *Western Political Quarterly* 2 (June 1976): 202–30.

112 **Even General Groves, the Manhattan Project**: "Chapter 8, Security Classification of Information," Federation of American Scientists, retrieved February 28, 2016, http:// www.fas.org/sgp/library/quist2/chap_8.html.

112 **The United States has invested over**: "Is a Career in Nanotechnology in Your Future?" National Nanotechnology Infrastructure Network, retrieved March 3, 2016, http:// www.nnin.org/news-events/spotlights/nanotechnology-careers.

112 **According to a 2016 report by**: Franz-Stefan Gady, "Top U.S. Spy Chief: China Still Successful in Cyber Espionage against U.S.," *Diplomat*, February 16, 2016, http://thediplomat.com/2016/02/top-us-spy-chief-china-still-successful-in-cyber-espionage-against-us.

113 **Stockholm International Peace Research Institute**: "16 June 2014: Nuclear Forces Reduced While Modernizations Continue, Says SIPRI," Stockholm International Peace Research Institute, http://www.sipri.org/media/pressreleases/2014 /nuclear_May_2014.

113 **The how-to information can be**: Tuan C. Nguyen, "Why It's So Hard to Make Nuclear Weapons," *Livescience*, September 22, 2009, , http://www.livescience .com/5752-hard-nuclear-weapons.html.

113 **Treaties, like the Nuclear Non-Proliferation Treaty**: "Section 7.0 Nuclear Weapon Nations and Arsenals," nuclearweaponarchive.org, August 2001, retrieved April 16, 2016, http://nuclearweaponarchive.org/Nwfaq/Nfaq7.html.

120 **Two studies of Islamic terrorists, one**: Alan Travis, "M15 Report Challenges Views on Terrorism in Britain," *Guardian*, August 20, 2008, http://www.theguardian .com/uk/2008/aug/20/uksecurity.terrorism1, retrieved March 2, 2016; and Olivier Roy, "What Is the Driving Force behind Jihadist Terrorism?" *Inside Story*, December 18, 2015, http://insidestory.org.au/what-is-the-driving-force-behind -jihadist-terrorism.

120 **Olivier Roy, a professor**: "Olivier Roy Interview (2007): Conversations with History; Institute of International Studies, UC Berkeley," retrieved March 2, 2016, http:// globetrotter.berkeley.edu/people7/Roy/roy07-con5.html.

120 **Afghan pathologist Yusef Yadgari's 2007 study**: Soraya Sarhaddi Nelson, "Disabled Often Carry Out Afghan Suicide Missions," NPR.org, http://www.npr.org/templates /story/story.php?storyId=15276485.

9. Fighting Fire with Fire

134 **Modern humans evolved about 200,000**: "Timeline: Weapons Technology," *New Scientist*, July 7, 2009, retrieved March 4, 2016, https://www.newscientist.com /article/dn17423-timeline-weapons-technology.

134 **Many experts estimate that the current**: Michael J. Mills et al., "Multidecadal Global Cooling and Unprecedented Ozone Loss Following a Regional Nuclear Conflict," *Earth's Future*, April 1, 2014, http://onlinelibrary.wiley.com/doi/10.1002/2013ef000205 /full.

134 **For example, the deployment of tanks**: Frederick Myatt, *Modern Small Arms* (New York: Crescent Books, 1978), 228–29.

134 **Even ICBMs are destroyable in flight**: "System A-135 Missile 51T6-ABM-4 GORGON," *Military Russia*, February 15, 2016, retrieved March 4, 2016, http://militaryrussia .ru/blog/topic-345.html; Ronald T. Kadish, "Reorganization of the Missile Defense Program," March 13, 2002, http://www.mda.mil/global/documents/pdf/ps_kadish 13mar02.pdf.

10. The Nanoweapons Superpowers

139 **Singularity Computers will first emerge**: Louis A. Del Monte, *The Artificial Intelligence Revolution* (North Charleston SC: Createspace, April 2014), 103–5, 128.

147 **According to a 2014 report published**: Michael J. Mills et al., "Multidecadal Global Cooling and Unprecedented Ozone Loss Following a Regional Nuclear Conflict," *Earth's Future*, April 1, 2014, http://onlinelibrary.wiley.com/doi/10.1002/2013ef000205/full.

149 **Recent experiments, performed in 2009**: Kristina Grifantini, "Robots 'Evolve' the Ability to Deceive," MIT Technology Review, August 18, 2009, https://www.technology review.com/s/414934/robots-evolve-the-ability-to-deceive.

11. The Nano Wars

153 **With the challenging international landscape today**: Frank Hoffman, "The Contemporary Spectrum of Conflict," *2016 Index of U.S. Military Strength*, retrieved March 15, 2016 http://index.heritage.org/military/2016/essays/contemporary-spectrum-of-conflict.

154 **We find ourselves in a spectrum**: Hoffman, "The Contemporary Spectrum of Conflict."

155 **How did we win?**: "How Were the Colonies Able to Win Independence?" *Digital History*, retrieved March 15, 2016, http://www.digitalhistory.uh.edu/disp_textbook .cfm?smtid=2&psid=3220.

157 **According to the Center for Naval**: Mary Ellen Connell and Ryan Evans, "Russia's 'Ambiguous Warfare' and Implications for the U.S. Marine Corps," *Center for Naval Analysis*, May 2015, https://www.cna.org/cna_files/pdf/dop-2015-u-010447-Final.pdf.

158 **We can define gray zone wars**: Hoffman, "The Contemporary Spectrum of Conflict."

158 **Nadia Schadlow made this observation**: Hoffman, "The Contemporary Spectrum of Conflict."

164 **Treaty on the Non-Proliferation of Nuclear**: "The Nuclear Non-Proliferation Treaty (NPT), 1968," U.S. Department of State Office of the Historian, retrieved March 15, 2016, https://history.state.gov/milestones/1961-1968/npt.

165 **Limited Test Ban Treaty**: "The Limited Test Ban Treaty, 1963," U.S. Department of State Office of the Historian, retrieved March 15, 2016, https://history.state.gov /milestones/1961-1968/limited-ban.

165 **Biological Weapons Convention**: "Convention on the Prohibition of the Development, Production, and Stockpiling of Bacteriological (Biological) and Toxin Weapons and on Their Destruction," United Nations, retrieved March 15, 2016, http:// www.un.org/disarmament/wmd/Bio/pdf/Text_of_the_Convention.pdf.

12. Humanity on the Brink

171 **Even though thirty-two people died**: "1986 Nuclear Disaster at Chernobyl," *History*, retrieved April 16, 2016, http://www.history.com/this-day-in-history/nuclear -disaster-at-chernobyl.

171 **An estimated 985,000 died**: Karl Grossman, "Chernobyl Death Toll: 985,000, Mostly from Cancer," *Global Research*, March 13, 2013, http://www.globalresearch .ca/new-book-concludes-chernobyl-death-toll-985-000-mostly-from-cancer/20908.

Epilogue

183 **The UN Commission for Conventional Armaments**: "A Study on Conventional Disarmament," United Nations, December 9, 1981, retrieved April 16, 2016, http://www .un.org/documents/ga/res/36/a36r097.htm.

186 **For these reasons and because of**: Ronald Reagan, "Foreword Written for a Report on the Strategic Defense Initiative, December 28, 1984," The American Presidency Project, http://www.presidency.ucsb.edu/ws/?pid=38499.

Appendix 2

193 **The Cientifica report, "Nanotechnology Funding**: Tim Harper, Cientifica Ltd., "Nanotechnology Funding: A Global Perspective," *Report of the National Nanotechnology Initiative Workshop*, May 1–2, 2012, Nano.gov, https://www.nano.gov/sites /default/files/pub_resource/nni_rsl_2012_rpt_0.pdf.

196 **Let's start by examining the 2015**: "The Military Balance 2016," International Institute of Strategic Studies, February 9, 2016, retrieved April 16, 2016, https://www .iiss.org/en/publications/military%20balance/issues/the-military-balance-2016-d6c9.

198 **In fact, military strategists characterize their**: James Wither, "An Endangered Partnership: The Anglo-American Defence Relationship in the Early Twenty-first Century," *European Security* 15, no. 1, March 2006, 47–65, retrieved April 16, 2016, http://www.tandfonline.com/doi/abs/10.1080/09662830600776694.

199 **They also are a close ally**: Robert Lacey, *Inside the Kingdom: Kings, Clerics, Modernists, Terrorists, and the Struggle for Saudi Arabia* (New York: Viking Press, 2009), 294.

201 **French nuclear weapons, like those**: "Status of World Nuclear Forces," Federation of American Scientists, retrieved February 29, 2016, http://fas.org/issues/nuclear-weapons/status-world-nuclear-forces.

201 **Some military analysts suggest that France**: Michael Shurkin and Peter A. Wilson, "France Is Replacing the UK as America's Top Ally in Europe," *Newsweek*, March 30, 2015, http://www.newsweek.com/france-replacing-uk-americas-top-ally-europe-317774.

201 **As of 2015, there are about**: "Nanotechnology," *Research in Germany*, retrieved March 2, 2016, http://www.research-in-germany.org/en/research-areas-a-z/nanotechnology.html.

201 **Historically Germany is a signatory**: "Statement of the Presidency of the Permanent Council of the WEU on behalf of the High Contracting Parties to the Modified Brussels Treaty—Belgium, France, Germany, Greece, Italy, Luxembourg, The Netherlands, Portugal, Spain, and the United Kingdom," *Western European Union*, Brussels, March 31, 2010, retrieved March 2, 2016, http://www.weu.int/Declaration_E.pdf.

201 **The Brussels treaty expired in 2010**: "Member Countries," NATO, retrieved March 2, 2016, http://www.nato.int/nato-welcome.

202 **South Korea is the world's fifteenth**: Kyle Mizokami, "It's Time for the U.S. Military to Leave South Korea," *The Week*, August 13, 2015, http://theweek.com/articles/570764/time-military-leave-south-korea.

Appendix 3

205 **At 1:06 a.m. (local time)**: Zhores Medvedev, *The Legacy of Chernobyl* (New York: W. W. Norton, 1990).

206 **In the parlance of nuclear physics**: Khalid Alnoaimi, "Xenon-135 Reactor Poisoning," March 15, 2014, http://large.stanford.edu/courses/2014/ph241/alnoaimi2.

GLOSSARY

ambiguous wars   Situations in which a state deploys troops and proxies in a deceptive and confusing manner, with the intent of achieving political and military goals, while obscuring the direct participation of the state or non-state.

analyte   Pertains to a substance being analyzed for its chemical composition.

artificial intelligence   Computers with software that allow them to emulate aspects of human intelligence.

autonomous weapons   Weapons, typically guided by artificial intelligence, that can act without human intervention.

biochemical   Pertaining to the chemical processes of living organisms.

Biological Weapons Convention (BWC)   The Convention on the Prohibition of the Development, Production, and Stockpiling of Bacteriological (Biological) and Toxin Weapons and on Their Destruction.

biosensor   A sensor with a biological component that aids in the detection of a substance.

boomer   Slang for a submarine that carries intercontinental ballistic missiles, typically to deliver nuclear weapons to a distant target.

brain implants   Technology implants that connect directly to the human brain to augment its intelligence and allow it to connect wirelessly with Singularity Computers.

catalyst   A substance that increases the rate of a chemical reaction.

centimeter   A unit of length in the metric system, equal to 1/100 of a meter.

cold war   Political hostility from 1945 to 1990 between the United States and Soviet Union characterized by threats, especially the threat of nuclear war, and propaganda.

counterinsurgency   Military or political action against guerrillas or revolutionaries.

defensive strategic nanoweapons   Defensive strategic weapons whose nanotechnology components enhance their strategic capabilities. It also includes defensive autonomous smart nanobots.

defensive tactical nanoweapons   Defensive weapons whose nanotechnology components enhance their tactical capabilities.

DNA   An acronym for deoxyribonucleic acid, a self-replicating material present in almost all living organisms and the carrier of genetic information.

doping   In integrated circuit fabrication, this refers to diffusing specific substances into unmasked portions of a substrate to change the electrical properties of that region.

electric field   A region around a charged particle, like an electron, or electrically charged object, like a "charged" balloon that children make by rubbing the balloon on their cloths to make it stick to a wall. Within the region, a force exerts on other charged particles or objects.

electromagnetic radiation   A type of radiation, such as visible light, gamma rays, X-rays, and radio waves, that includes simultaneously varying electric and magnetic fields.

electron   A fundamental subatomic particle with a negative charge found in atoms.

electron beam lithography   A maskless technology in which a pattern results from a digital representation on computer, which controls scanning an electron beam across an electron sensitive resist-coated substrate.

electrons scanning microscope   A microscope that uses a beam of electrons to produce images of a sample.

emtech exploitation index   A measure regarding how effectively a country can commercially exploit its nanotechnology investments.

environmental obfuscation   Pertaining to the use of obscurants, like white smoke or strobe lights that make it difficult to determine elements in an environment.

Eugene Goostman   A software program that emulates the persona of a thirteen-year-old Ukrainian boy.

European Framework Programmes for Research and Technological Development   Also termed "Framework Programmes" or abbreviated "FP1 through FP7," with "FP8" being named "Horizon 2020," are funding programs by the European Union/European Commission to support research in the European Research Area (ERA).

father of all bombs   The nickname of Russia's most powerful nonnuclear air-delivered bomb, speculated to use nanotechnology, which makes it four times more destructive than the U.S. Massive Ordnance Air Blast bomb, nicknamed the "mother of all bombs."

ferroelectric   A substance that displays a permanent electric polarization in proportion to an applied electric field.

field-effect transistor   A transistor in which current flows across a channel controlled by a transverse electric field.

full-out nano war   A nano war that includes the use of strategic nanoweapons.

gels   A substance with the constancy of jelly.

gene therapies   Transplantation of genes into cells in order to correct genetic disorders or alter genetic programming.

graphene   A single layer of carbon atoms structured like a honeycomb, with excellent mechanical strength and electrical conduction properties.

gray zone wars   Deliberate activities by a nation, or fractions within a nation, that seek to achieve strategic goals without the use of military forces.

GRU   An acronym for Glavnoye Razvedyvatel'noye Upravleniye, which is Russia's largest foreign intelligence agency.

helium leak detector   An apparatus that detects helium gas.

hybrid wars   Any blend of regular and irregular tactics involving a violent struggle among state and non-state actors for legitimacy and influence over the relevant populations.

hydrophilic   The tendency of a substance to mix with, dissolve in, or be "wetted" by water.

ICBM   An acronym for "intercontinental ballistic missile."

imaging sensor   A device that converts an optical or infrared image into an electric signal.

integrated circuits   Solid-state silicon "chips" (integrated circuit chips) that embody electronic circuits and, potentially, sensors.

integrated circuit chip   Typically a small, thin, flat rectangular or square piece of silicon that contains numerous electronic and/or sensor elements.

intelligent agent   A microprocessor running a program that enables it to perform functions equivalent to a human expert performing the same function.

Intercontinental Ballistic Missile (ICBM)   A ballistic missile with a minimum range greater than 3,400 miles and designed to deliver nuclear weapons.

International Humanitarian Law   A set of rules and principles that requires nations to act "humanely" during war.

limited conventional wars   Conflicts between nations using conventional military means, limited by geographic boundaries, types of targets, and disciplined use of force.

Limited Test Ban Treaty (LTBT)   A treaty that prohibits the testing of nuclear weapons in the atmosphere, under water, or in space.

lipid   A class of fatty acid organic compounds that is insoluble in water but soluble in organic solvents.

lyophilized medicines   Freeze-dried medicines.

mach   Ratio of the speed of a body relative to the speed of sound (e.g., Mach 10 = 10 times the speed of sound, where the speed of sound = 1,125 feet per second).

magnetotactic bacteria Bacteria that migrate along the Earth's magnetic field lines.

meta-materials A synthetic composite material that exhibits properties not found in natural materials.

meter A unit of length in the metric system, equal to 100 centimeters or approximately 39.37 inches.

micelle A collection of molecules in a colloidal solution (i.e., floating in a solution).

microprocessor A computer that resides on an integrated circuit chip.

mini-nuke A nuclear bomb briefcase-sized or smaller made possible by nanotechnology. Mini-nukes are purportedly under development by the United States, Russia, and Germany. There is speculation that the nuclear device utilizes a nano laser capable of triggering a comparatively small thermonuclear explosion when directed into a mixture of tritium and deuterium.

Moore's law An observation by Intel founder Gordon Moore that the density of integrated circuits doubles every two years in a cost-effective manner.

nano war Any war that involves nanoweapons.

nanoantenna A nanoscale antenna for transmitting electromagnetic waves, such as visible light.

nanobubble A bubble with a diameter at the nanoscale, 1–100 nanometers.

nanocrystalline A nanoscale polycrystalline material.

nanometer A unit of length in the metric system, equal to one-billionth of a meter.

nanotechnology Science, engineering, and technology conducted at the nanoscale, which is 1–100 nanometers.

nanotechnology impact factor A metric that measures how the impact of a nation's thrusts in nanotechnology is likely to affect its economy.

nanotube A tubular carbon molecule.

nanoweapon Any military technology that exploits the use of nanotechnology.

nanoweapons capable nations   Nations capable of developing nanoweapons that have chosen not to do so.

nanoweapons nations   Nations already deploying nanoweapons and having robust ongoing nanoweapons development programs.

nanoweapons power   Nations with nanoweapons and supercomputers that are one or more generations behind nanoweapon superpowers.

nanoweapons superpowers   Nations that have Singularity Computers and the best in class nanoweapons, especially self-replicating smart nanobots (SSN), programmable to perform numerous functions.

nanowire   A semiconductor nanoscale rod used in some transistor and laser applications.

National Nanotechnology Initiative (NNI)   According to http://www.nano.gov/about-nni/what, "a U.S. government research and development (R&D) initiative involving the nanotechnology-related activities of 20 departments and independent agencies."

NATO   An acronym for North Atlantic Treaty Organization.

near follower nanoweapons nations   Nations developing or deploying nanoweapons that are one or more generations behind nanoweapons nations.

NIF   An acronym for Nanotechnology Impact Factor.

NOCON   An acronym for Nanoweapons Offensive Capability of Nations.

NOCON score   A metric that expresses a nation's capability to develop and deploy offensive nanoweapons, which is defined by the following equation: NOCON score = (military budget) × (NIF expressed as a decimal).

NORAD   An acronym for North American Aerospace Defense Command.

Non-Proliferation Treaty (also known as the Treaty on the Non-Proliferation of Nuclear Weapons)   An international treaty to prevent the spread of nuclear weapons, to promote cooperation in the peaceful uses of nuclear energy, and to achieve nuclear disarmament and general and complete disarmament.

North American Aerospace Defense Command   An organization of the United States and Canada whose mission is to provide aerospace warning, air sovereignty, and defense for North America.

North Atlantic Treaty Organization   A military alliance of European and North American democracies to strengthen international ties between member states and historically to counterbalance the Warsaw Pact.

obfuscation   Obscuring elements in the environment.

obscurant   A material or method to darken, confuse, stupefy, or bewilder.

offensive strategic nanoweapons   Weapons whose nanotechnology components enhance their strategic capabilities. It also includes (hypothetical) offensive autonomous smart nanobots.

offensive tactical nanoweapons   Offensive weapons whose nanotechnology components enhance their tactical capabilities.

optoelectronic   Technology that focuses on the combination of electronics and light.

passive nanoweapons   Any use of nanotechnology in warfare that has a nonoffensive/nondefensive application, but may increase the effectiveness of conventional or strategic weapons.

phonon   A quantum of energy associated with a sound wave or vibration in a crystal lattice.

photon   A zero mass quantum of electromagnetic radiation, including visible light.

photonic   Pertaining to photons.

photonic crystals   A synthetic optical nanostructure that affects the motion of photons.

plasma   An ionized gas consisting of equal proportions of positive ions and free electrons resulting in no overall electric charge.

plasmonic   A quantum of plasma oscillation.

polycrystalline   Consisting of many randomly oriented crystalline parts.

processing power   The speed at which a computer can perform an operation.

purchasing power parity   The adjustment needed regarding the exchange rate between countries for the exchange to be equivalent.

quantum dots   Semiconductor nanoparticles.

quantum mechanics   A theory in physics that holds all matter at the level of atoms and subatomic particles possess a dual wave/particle nature, mathematically interpreted as "wavefunction," which completely describes the atom or subatomic particle's physical state and its change with time.

radiation   The emission of electromagnetic waves or subatomic particles.

resist   A photo or electron sensitive material that when exposed to photons or electrons, and appropriately processed, can act as an etching mask.

RFID   An acronym for radio frequency identification, which is a technology that uses electromagnetic or electrostatic coupling in the radio frequency range of the electromagnetic spectrum to identify an object, animal, or person.

scanning tunneling microscope   A type of microscope that works by detecting electrical forces with an ultrafine tip probe to reveal the atomic and molecular surface of a sample.

Schrödinger equation   An equation that forms the basis for the mathematical description of atoms and subatomic particles in quantum mechanics.

self-replicating smart nanobots (SSN)   Hypothetical smart nanobots that are able to replicate by building copies of themselves from raw materials, similar to how a biological virus reproduces.

Singularity Computers   Computers that are more intelligent than the combined cognitive intelligence of humanity.

smart   An adjective often used to modify a product having artificial intelligence, such as "smart phone."

smart nanobots   Nanobots with artificial intelligence.

smart prosthetics   Prosthetics that embody artificial intelligence that enable it to connect directly to the central nervous system and perform all normal functions, equivalent to the limb it replaced.

spectrum of conflict   The different forms of conflict, from full-out nuclear "theater conflict" to provocative behavior "gray zone" war.

spin physics   In the physics of quantum mechanics, "spin" relates to the intrinsic angular momentum of elementary particles, composite particles, and atomic nuclei.

spintronics   The study of electron spin (i.e., a component of it angular momentum).

SSN   An acronym for self-replicating smart nanobots

stability operations   Defined by DoD Instruction 3000.05 as "various military missions, tasks, and activities conducted outside the United States in coordination with other instruments of national power to maintain or reestablish a safe and secure environment, provide essential governmental services, emergency infrastructure reconstruction, and humanitarian relief."

subatomic particle   Any particle that is required to form an atom, such as an electron.

substrate   In integrated circuit manufacture, this structure embodies the individual integrated circuit chips prior to being "diced" (cut apart).

superthermite   Also known as super-thermite and nano-thermite, refers to a mixture of nanoparticles, oxidizers, and reducing agents, such as iron oxide and aluminum nanoparticles, which react more rapidly than their micro counterparts do because of the high surface to volume ratio of nanoparticles.

Technological Singularity   A specific point in time when technology profoundly alters the path of human evolution.

thermite A mixture of powdered aluminum and iron oxide, which produces a high temperature during combustion, making it ideal for incendiary bombs.

Treaty on the Non-Proliferation of Nuclear Weapons According to the United Nations Office for Disarmament Affairs, this is an "international treaty whose objective is to prevent the spread of nuclear weapons and weapons technology, to promote cooperation in the peaceful uses of nuclear energy, and to further the goal of achieving nuclear disarmament and general and complete disarmament."

Warsaw Pact A military alliance of communist nations in Eastern Europe, organized in 1955 to counter NATO but dissolved in 1991.

wavefunction In quantum mechanics it pertains to a mathematical object, typically designated by the Greek letter ψ, that represents the quantum state (the most complete physical description) of a specific isolated system of one or more particles.

WU-14 The Pentagon's code name for a Chinese experimental hypersonic glide vehicle (HGV), now called the DF-ZF.

INDEX

abalone shells, 17, 30
accountability and autonomous weapons, 105–6
active nanostructures, 31
AFM. *See* atomic force microscope (AFM)
aging therapies, 32, 140
AI. *See* artificial intelligence (AI)
alliances, national: American Revolution and, 155, 156; nanotechnology capabilities and, 114–15, 116, 144, 145; new cold war and, 146, 147, 198
ambiguous wars, 157–58
American Revolution, 155–57
Ancestry.com, 87
Andraka, Jack, 86
antiballistic missile defense systems, 72, 163, 185, 186
antisatellite missiles, 70–71
Anti-Submarine Warfare Continuous Trail Vessel, 161
antivirals, 135, 139
armor: body, 55, 56–57, 72, 160; tank, 30, 55, 59, 134
arms race, nanoweapons, xii, 5–6; major players in, 64–67; new cold war and, 148, 151, 183; secrecy of, 111; treaties and, 165; vicious circle of, 136. *See also* nanoweapons, foreign development of
arms race, nuclear, 181
Army Research Laboratory, 85
artificial intelligence (AI), 225; development of, 61–62, 104–5, 107, 136; equality of human intelligence to, 61–62, 102–7; exceeding human intelligence, 110, 118, 139, 149, 151; functions in, 101–

2; general, 103; nanobots and, xxiii, 6, 86, 88, 92–94, 97, 105, 108; software programs and, 102; term use of, 101–2
artillery, 56, 59–60, 160–61
Asimov, Isaac, 151
ASNS. *See* autonomous smart nanobots (ASNS)
asymmetrical warfare, 71, 73–74, 157
atomic bombs, 13, 111–12, 165, 183
atomic control, 21, 24
atomic force microscope (AFM), 24
atomic layer epitaxy, 38
atomic submarines, xiv
atoms, xiii, 10, 17, 19, 20–22, 24
attacks: autonomous weapons and, 106–7, 108–10; cyber, 70; detecting and determining source of, 70, 119, 167; nanoweapons scenario, 3–6, 75–76, 109–10, 125–33, 174–77; nuclear *versus* nano, 108; swarming, 91–97
Autonomous Rotorcraft Sniper System, 84, 161
autonomous smart nanobots (ASNS), 108–10, 135, 162–63, 177, 185–86, 187
autonomous weapons, 79, 225; ethical considerations for, 106–7; nano, 87, 107–10, 135, 162–63, 177, 185–86, 187; non-nano, 84–85, 105–6, 161. *See also specific types*

bacteria, 41, 135
balance of power, 75, 107
batteries, 52–53, 57
Binnig, Gerd, 21
biological life-forms, 184

biological weapons, xi, xiii, 43–44, 165–66, 182, 183
Biological Weapons Convention (BWC), 44, 165–66, 184–85, 187
biosensors, 24, 40, 48, 160, 225
black market, 121
body armor, 55, 56–57, 72, 160
boomers: about, 93, 95, 108, 125; in nanoweapon scenario, 125–26, 131–33
boost-glide technology, 63, 69–70
botulinum toxin type H, 6, 7
Bradley, Omar N., 45
brain implants, 140, 225
bridge collapses, 36–37
Brynjolfsson, Erik, 101
budgets: military, 69, 117–18; nanotechnology, xii, 8–9, 23, 35, 65–67
bullets, smart, 55, 58–59, 160
burn victims, treatment for, 41
BWC. See Biological Weapons Convention (BWC)

cancer nanomedicine, 41–42, 86–87, 88
Center for Naval Analysis, 157
Centre National de la Recherché Scientifique, 115–16, 144
ceramics, 52–53
chemical nanosensors, 48–49, 160
chemical weapons, 166
Chernobyl nuclear disaster, 169–72, 178, 181, 205–9
China, 113, 146, 158, 165; assymetric warfare capabilities of, 73–74; autonomous weapons and, 84, 106; espionage and, 70, 112–13, 142; nanotechnology development and, 5–6, 23, 62–64, 69–72, 114–15, 119, 142–43; nanotechnology spending of, 65, 66–67, 194–96; Russian trade with, 71-72; scenarios and, 75–76, 127–30; submarines and missiles and, 95, 186; supercomputers and, 102–3
chips, computer, 38–39, 48, 104
chromallocytes, 43
Chubais, Anatoly, 68
Churchill, Winston, 188

Cientifica, 66–67, 193, 195–96, 197, 198
Clapper, James R., 112–13
Clark, Arthur C., 184
classified information, xii, 8–9, 45–46, 112
coatings, 37–38, 47–48, 52–53, 58–59, 160, 161–62
Cold War, 72, 95, 105, 148, 164
cold wars, future, 119–21, 146, 148–49, 151–52, 167, 185
collective self-defense, 164–67
commercial uses of nanotechnology. See nanotechnology, commercial uses of
computer-aided design of nanoweapons, 107–8
computer processing power, 102–4
concrete, 36, 160
construction, 30, 36–38
consumer nanotechnology products, 30, 32–36, 49
Convention on the Prohibition of the Development, Production, and Stockpiling of Bacteriological (Biological) and Toxin Weapons and on Their Destruction, 44, 165–66, 184–85, 187
corrosion, 37–38, 52–53, 89, 160
counterinsurgency and stability operations, 154–57
Crimean War, 157–58
critical events for emergence of nanotechnology, 21–22
cyber espionage, 70, 112–13
cyber warfare, 70

Defense Advanced Research Projects Agency (DARPA), 8, 58, 82, 84–85, 87
"Defense Nanotechnology Research and Development Program" (DoD), 46, 50–52, 60–61
Defense One, 84, 106
Defense Threat Reduction Agency, 40
defensive nanoweapons, 134–35; scenario on, 123–33; strategic, 163; tactical, 161–62, 164
defensive strategic nanoweapons, 163
defensive tactical weapons, 161, 164

Denny, Reginald, 81–82

Dennymites, 81–82

Department of Defense (DoD): nanoweapons conferences and, 40; NNI and, 8–9; research and development and, 11, 46, 47–48, 50–52, 60–61, 63, 159

destruction capability growth, 134

detection: of manufacture of nanoweapons, 50; of nanoweapons, xiii, 13, 14, 48–49, 108, 119; of nucl`ear missiles, 108

deterrence of war, 74, 146, 186

development of nanotechnology. *See* research and development of nanotechnology

dip pen nanolithography (DPN), 24, 25

DN-2S, 70–71

DN-3S, 71

DNA: human nano processes and, 19; nanobots and, 85–86, 87–88, 94, 96, 97; replication of, 140; smart bullets and, 58

DPN (dip pen nanolithography), 24, 25

Drexler, Kim Eric, 10–11, 21, 22

drones, 61–62, 79, 82; autonomous, 105, 106–7; evolution of, 81–85; future military development of, 161–62; micro and nano, 87–88, 161. *See also* Sopwith ATS; Wickersham Land Torpedoes

drug delivery, 20, 41, 42, 85–86

Earth's Future, 147

economic wars, 148

ecosystem, Earth's, 147

Eigler, Don, 21, 24

electric lasers, 59

electromigration, 31

electron beam lithography, 38–39

Eleven General Orders, 93

embedded sensors, 56–57

Emtech Exploitation Index (EEI), 195

energy for Singularity Computers, 149

Engines of Creation (Drexler), 10, 21, 22

Environmental Protection Agency (EPA), 35

envisionment of year 2050. *See* year 2050 envisionment

espionage, 70, 72, 73, 111–13, 142, 148

European Framework Programmes for Research and Technological Development, 23, 115–16, 144, 145

European Medicines Agency, 41

European Patent Office, 67

European Union (EU), 23, 65, 66

exoskeletons, 57

explosives, 11, 47–48, 55–56, 59–60, 160, 161

extinction of humanity, xi–xiii, 75–76; cold war risk of, 146, 147–48, 149, 152; nanobots and, 7, 140; nuclear arsenal and, 134; prevention of, 164, 185–86

Faculty of Life Sciences, 95–96

fail-safe override of Singularity Computers, 178

fallout, radioactive, 13, 63–64, 165, 171, 179

Fast Lightweight Autonomy program (FLA), 84–85

"father of all bombs," 11, 13, 66

fatigue of steel, 36–38

Feynman, Richard, 10–11, 17

flowers, 19, 92

Food and Drug Administration (FDA), 35

foreign development of nanoweapons, 64–77. *See also* "Nanoweapons Offensive Capability of Nations" (NOCON); *specific nations*

fourth general order, 93

Framework Programmes, 23, 115–16, 144, 145

France: American Revolution and, 156; Limited Test Ban Treaty and, 165; nanoweapons and, 114, 115–16, 144, 195, 197; nuclear weapons and, 113, 165; terrorism and, 120

Franklin, Benjamin, 112

funding: of the military, 69, 117–18; of nanotechnology, xii, 8–9, 23, 35, 65–67

future cold wars, 119–21, 146, 148–49, 151–52, 167, 185

General Atomics, 83

gene therapies, 87, 140

genetic testing, 87
Geneva Protocol, 165–66
genocide, 166, 175–77
Germany: arms race and, 6, 12–13; nanoweapons development and, xxii, 12, 73, 116, 144–45, 195–98; World War I and II weaponry of, 81–82, 134
Global Catastrophic Risk Conference, xi
Global Positioning System (GPS), 83
gold, 26
Goliath track mines, 81
Goostman software program, 103, 104
graphene, 54
gray zone wars, 158–59
Great Britain, 155–57
Groves, L. R., 111, 112
guerrilla tactics, 155–56

H1N1 virus, 92
Health and Consumer Protection Directorate, 35
Health and Fitness nanotechnology products, 33, 34–35
helicopters, unmanned, 84, 161
Hong Kong, 143
hotline, Moscow-Washington, 123
human brain processing capability, 102–3
human extinction. *See* extinction of humanity
human intelligence: computer intelligence equal to, 61–62, 97, 102–7; computer intelligence exceeding, 110, 115, 139, 149, 151
humanitarian law, international, 106
hybrid wars, 157
hydrogen bombs, 165
hypersonic missiles, 62–63, 69–70, 75, 108, 162–63, 185

image sensors, 140, 160
imaging systems, 60, 61
India: nanoweapons capability of, 117, 194–98; nuclear weapons and, 113, 164
industrial nanotechnology products, 30, 36–40, 49, 160
informed population, 165–66, 181–83
Institute for Nanoscience, 50–51, 52

Institute for Soldier Nanotechnologies (ISN), 55–56, 189–92
Institute of Nanotechnology & Advanced Materials, 95–96
integrated circuits, 8–9, 38–39, 42, 46–47, 48, 49, 104, 159–60
Intel, 34, 71
intelligence and computers, 61–62, 97, 102–7, 110, 139, 151, 178
intelligence augmentation, 140
intelligent agents, 102–3
intercontinental ballistic missiles (ICBMS), 75, 186
intercontinental hypersonic missiles, 62–64, 69–70, 72, 75
International Symposium on Assessing the Economic Impact of Nanotechnology, 29–30
invisibility cloaks, 55, 57–58, 160, 162
Iraq, 84, 118, 166
irregular warfare, 157
ISN. *See* Institute for Soldier Nanotechnologies (ISN)
Israel, 113, 164

Japan, 37, 113; atomic bombings in, 13, 111, 165; nanotechnology spending of, 65, 66; nanoweapons capability of, 117, 146, 194–98; in scenarios, 174
judgment, human, 105–6, 107, 172

Kettering Bugs, 80
Kidlington Centre, 42
Kurzweil, Ray, 22, 110

lab-on-a-chip devices, 48
Laboratory of Intelligent Systems experiment, 149–50
large theater conflicts, 154
laser weapons, 11–13, 12, 39, 56, 59, 64, 152, 160
Lausanne experiment, 149–50, 151
Lawrence Berkeley National Laboratory, 57
leaks, information, 112–13
life-forms, 149, 184, 187
Li-ion batteries, 57

238

Limited Test Ban Treaty (LTBT), 165, 185–86, 187
Lockheed, 9, 83
Lomasney, Christina, 59
lotus effect, 19

MAD (mutually assured destruction), 43, 74, 95, 109, 146, 147, 186
magnetotactic bacteria, 135
Manhattan Project, 15, 112
marine growth, 53, 160
Markov, Georgi, 6–7
Massive Ordnance Air Blast bombs, 11
McGinnis, Sean P., 33
medical nanotechnology. *See* nanomedicine
metal: electromigration and, 31; nano-enhanced, 11, 47, 58–59, 161–62; nanoscale and, 26; silver, 33, 41; steel, 30, 36–38, 58, 160, 161–62
micelles, 18, 20
micro drones, *82, 85, 87*
The Military Balance, 117–18, 196, 199
mini-nukes, 12–13, 45, 63–64, 73, 116, 161
missile defense systems, 72, 106–7, 163, 185, 186
missiles, 62–63, 69–71, 75, 81, 108, 162–63, 185. *See also specific types*
MIT *Technology Review,* 38, 48, 73
Mobile Autonomous Robot Software program, 84
mobile lasers, 59
molecular manipulation, 10, 17
molecular manufacturing, 22
molecular self-assembly, 25
monetary value, redefinition of, 149
Moore, Gordon, 104
Moore's law, 104, 141
Morocco, 156
mosquito-like nanobots, 6, 7
"mother of all bombs," 11
MQM-105 Aquilas, 83
munitions, nano-enhanced, 61
mussels, 17–18
mutually assured destruction (MAD), 43, 74, 95, 109, 146, 147, 186

mutually assured extinction, 147–48, 152

"nano," word of, 35
nanobiosensors, 48
nanobiotechnology, 25
nanobots, xiii; autonomous, artificially intelligent, and self-replicating, 7–8, 108–10, 140, 142, 162, 163, 184–86, 187; defensive smart, 134–35, 163; evolution of, 79–85, 88; expectations of, 87–89, 148; mosquito-like, 6, 7; nanomedicine and, 42–43, 85–87; in scenarios, 74, 75–77, 88, 123–33, 174–77; swarming, 92–97
nanobubbles, 18
nanoceramic coatings, 52–53
nanoelectromechanical systems, 25
nanoelectronic integrated circuits, 46–47, 159–60
nanoelectronic processors, 63, 70, 71, 140
nanoenergetics, 59
nanofactories, 88–89
nanolithography, 24, 25, 49
nanomanufacturing, 38–40, 43
nanomaterials, defined, 29
nanomedicine: commercial use of, 40–43, 96; military use of, 47, 57, 160; nanobots and, 85–87, 96–97, 135; Singularity Computers and, 139–40
nanoparticles, 13, 25, 26; commercial nanotechnology applications and, 29, 31, 35, 36, 37, 41; manufacture of, 49; medicine and, 41, 86; in nature, 17, 19; toxicity and health hazard of, 13–15, 22–23, 35, 41; weaponry and, 47, 49, 56, 59, 60, 63, 70, 160
nano processes, 17–20
nanorobotics, 40, 42–43, 49, 161. *See also* nanobots
nanoscale, 21–22, 25–26, 30–31, 39
nanoscale proteins, 19–20
"Nanoscience and Nanotechnologies" (Royal Society and Royal Academy of Engineering) report, 22–23
nanosensors, 48–49, 60, 160
nanosilver-impregnated bandages, 160

nanosystems, 31–32

Nanosystems (Drexler), 10

Nanotech Impact Factor, 66–67

nanotechnology: approaches to building in, 24–25, 38, 51; budgets, xii, 8–9, 23, 35, 65–67; categories of, 29–30, 31–32; complexity of, 172, 173; control of, xiii, 178; as enablers, 30, 49; environmental impact of, 35–36, 41; molecular, xi, 32; nanobots and, 88, 97, 135; Singularity Computers and, 151, 167, 173; defined, 21–22, 29–30, 159; environmental impact of, 35–36, 41; interest and market for, 22–23, 32, 42, 121; market for, 32, 42, 121; natural, 17–20; overview of military use of, 43–44; products, classifying of, 29–30, 31–32; regulation of, xii, 14, 23, 35, 50, 142; spending for, xii, 8–9, 23, 35, 65–67, 112. *See also* nanotechnology, commercial uses of; nanotechnology, control of; research and development of nanotechnology; *specific aspects of*

nanotechnology, commercial uses of, 23; consumer products, 30, 32–36, 49; industrial products, 30, 36–40, 49, 160; in manufacturing, 38–40; medical, 40–43, 96; overview of, 29–32; product categories, 29–30, 31–32; transfer to nanoweaponry, 46–47, 49, 65, 160

Nanotechnology Consumer Products Inventory (cpi), 32–33

"Nanotechnology Funding: A Global Perspective" (Harper), 66, 193, 195, 196, 198

Nanotechnology Impact Factor (nif), 193–94, 195–98

nano war, 159–66

nanoweapons: attack scenarios with, 3–6, 75–76, 108–10, 125–33, 174–77; awareness of, 8–9, 10, 13, 166, 181–82; categories of, 159–63; cloak of secrecy about, xiv, 10, 40, 45–46, 64, 148, 182; countermeasures to, 134–36; defined, 53, 159; design of, 107–8, 110, 148–49; ethical issues of, 43, 106–7; detection

of, xiii, 13, 14, 48–49, 108, 119; foreign development of, 64–77; manufacturing of, 35, 49–50, 51, 75, 119, 121, 136; nuclear weapons safety compared to, 147–48; as the ultimate threat, 178; use of to increase awareness of, 10, 173–77; warfare application of, 163–64; as weapons of choice, 118–19. *See also* arms race, nanoweapons; nations and nanoweapons capability; *specific aspects and types of*

"Nanoweapons Offensive Capability of Nations" (nocon), 113–18, 141–42, 193–203; conclusions and rationale of, 200–203; methodology of, 193–98; results of, 198–200; summary of, 113–18

nanowerk.com, xii, 54

Nanowerk News, 57–58

nasturtium flowers, 19

National Nanotechnology Initiative (nni), xii, 8, 21–23, 35, 38–40, 46, 65–66, 193

National Nanotechnology Initiative Workshop, 193

nations and nanoweapons capability, 113–18, 193–203. *See also specific nations*

nato, 201; Kosovo conflict and, 83; nuclear attack and, 108; superpower projections and, 143, 144, 145, 146, 147

natural nanotechnology, 17–20

"Naval Science & Technology Strategy," 52

new arms race. *See* arms race, nanoweapons

new cold wars, 119–21, 146, 148–49, 151–52, 167, 185

Next Big Future, 85–86

Nixon, Richard, 43–44

nni. *See* National Nanotechnology Initiative (nni)

nocon ("Nanoweapons Offensive Capability of Nations"). *See* "Nanoweapons Offensive Capability of Nations" (nocon)

Non-Proliferation Treaty, 113, 116, 144, 164–65, 185, 187

North Korea, 75–76, 113, 114, 117, 119, 146, 164, 186

Northrop Grumman, 59, 83
nuclear attacks, 108
nuclear fallout, 13, 63–64, 165, 171, 179
Nuclear Non-Proliferation Treaty, 113, 116, 144, 164–65, 185, 187
nuclear weapons, 43, 63, 75, 76, 93; detonation of, 147–48; information on, 10, 113; manufacturing of, 49–50, 136; nanoweapons compared to, 108–9, 119, 121, 179; nations with, 113; Non-Proliferation Treaty on, 113, 116, 144, 164, 185, 187; Singularity Computers and, 141, 147. *See also specific types*

offensive strategic nanoweapons, 162–64, 167. *See also specific types*
offensive tactical nanoweapons, 160–61, 163–64. *See also specific types*
Office of Naval Research, 52, 53, 54
oil, 118, 120, 145
Oppenheimer, Andy, 73
OQ-1s, 81–82
organs, artificial, 140
override of Singularity Computers, 178
oxygen, 19, 20, 94

Pakistan, 113, 118, 148, 164
passive nanomaterials, 56
passive nanostructures, 31
passive nanoweapons, 159–60, 163–64
patents, 32, 37, 65, 67
peace, 153
Persian Gulf, 118
Persian Gulf War, 83
Pfizer, 85–86
pharmaceuticals, 139
phone link between U.S. and Russia, 123
political interest in nanotechnology, 23
Ponce, USS, 11–12, 12
portability of nanoweapons, 121
power, self-generating, 109
power control systems, 53–54
PPP (purchasing power parity), 194–95
prevention of extinction of humanity strategies, 164, 185–86
processing capability of human brain, 102–3

processing power, computer, 102–4
profiles of terrorists, 120–21
projectiles, 48, 55–56, 58, 59–60
Project on Emerging Nanotechnologies, 32
proliferation of nanoweapons, 119
proliferation of nuclear weapons, 113, 116, 144, 164, 185, 187
propaganda, 148
prosthetics, smart, 140
proteins, nanoscale, 19–20
public, informed, 165, 166, 181–83
public information on nanoweapons, 9–11
Purchasing Power Parity (PPP), 66, 194–95
Putin, Vladimir, 68, 72, 158

quantum mechanics, 21, 25–26, 31
quantum nanoscience, 26

RA-115s, 75
radiation-hardened integrated circuits, 8–9, 47, 49
radioactive fallout, 13, 63–64, 165, 171, 179
raw materials for Singularity Computers, 149
regulation of nanotechnology, xii, 14, 23, 35, 50, 142
religious unification against nanoweapons, 146
remote controlled vehicles, 79, 80–83, 84. *See also* drones
research and development of nanotechnology: artificial intelligence and, 61–62, 102–5, 107, 136; Department of Defense and, 11, 46, 47–48, 50–52, 60–63, 159; nanomedicine and, 86–87; National Nanotechnology Initiative and, xii, 8, 21–23, 35, 38–40, 46, 65–66; overview of, xii–xiii, xiv, 22–24, 26–27; Singularity Computers and, 141; U.S. Air Force and, 60–62; U.S. Army and, 55–60, 85, 189–92; U.S. Navy and, 50–54
research category of nanotechnology, 22–24
resists, 25, 38–39

revolutionary nature of nanotechnology, 30–31, 38

robotic drone, 81. *See also* Dennynmites; Goliath track mines; Kettering Bugs; OQ-1S; VB-1 Azons, MQM-105 Aquilas; RQ-1 Predators; RQ-4 Global Hawks; *Sprengbootes*

Rogozin, Dmitry, 72

Rohrer, Heinrich, 21

Royal Academy of Engineering, 22

Royal Society, 22, 103

RQ-1 Predators, 83–84

RQ-4 Global Hawks, 83–84

"Runaround" (Asimov), 151

Rusnano, 67–69

Russia, 95, 105, 113, 158, 185; arms race and, 5–6; autonomous weapons and, 84, 106; Chinese trade with, 71–72; Crimean war and, 157–58; future cold wars and, 119–20, 146, 147; nanotechnology development and, 11, 12–13, 23, 62–64, 71–75, 84; nanotechnology spending of, 65–67, 194–95; as a nanoweapons power, 115, 142, 145; Rusnano and, 67–69; in scenarios, 76, 106–7, 123–33. *See also* Soviet Union

Russian Corporation of Nanotechnologies, 23, 67–68. *See also* Rusnano

Russian Strategic Missile Forces, 84, 106

satellites, 70–71, 108, 136, 148

Saudi Arabia, 117–18, 145

Sause, Brittany, 48

SC-19S, 70

scanning tunneling microscopes (STM), 21, 24

scenarios of nanoweapons attacks, 3–6, 75–76, 109–10, 125–33, 174–77

Scientific American, 11

secrecy of nanoweapons, xiv, 10, 40, 45–46, 64, 148, 182

self-connecting stem cells, 140

self-powered nanobots, 109

self preservation of nanotechnology, 149–50, 151

self-replicating smart nanobots (SSN), xiii, 8, 110, 140–42, 162, 163, 184–85, 187

self-steering bullets, 58

sensors, 24, 40, 48–49, 51, 56–57, 60, 140, 160

S-400 Missile Defense Systems, 72

silicon on insulator (SOI), 47

silver, 33, 41

singularity, 110, 139, 140

Singularity Computers: capabilities of, 139–41; China and, 143; control of, 167, 173, 178; development of, 148–49; disruptions of, 141–46; interfacing of, 143–44, 151; military prowess and, 150–51; raw materials and energy for, 149; Russia and, 145; self preservation of, 149–50

size and nanotechnologies, xiii, 6, 22, 26, 30–31, 73

smart artillery, 59–60, 160–61

smart bombs, 101, 105

smart bullets, 55, 58–59, 160

smart drugs, 41

smart nanomaterials, 56

smart phones, 101

smart projectiles, 56, 60

smart prosthetics, 140

smart robots, 105–6. *See also* nanobots

smuggling, 74–75

snipers, 55, 84, 160, 161

software, 84, 102–3, 104

SOI. *See* silicon on insulator (SOI)

Sopwith ATS, 80

South Korea, 65, 117, 119, 195–98

South Sudan, 164

Soviet Union, 80–81, 111, 164, 165

Spain, 156

Spanish flu pandemic, 92

spectrum of conflict, 153–59

Sprengbootes, 80

SSNS (self-replicating smart nanobots), xiii, 8, 110, 140–42, 162, 163, 184–85, 187

stealth aircraft, 9

stealth destroyers, 54

stealth weapons, 62–63, 108, 141, 143

steel, 30, 36–38, 49, 58, 160, 161–62

stem cell repair, 140

STM (scanning tunneling microscopes), 21, 24

Stockholm International Peace Research Institute, 113

Strachan, Hew, 153

Strategic Command, 62

strategic research areas (SRAS), 55, 189–92

"Studies on Nano-additive for the Substitution of Hazardous Chemical Substances in Antifouling Coatings for the Protection of Ship Hulls" (X. Zhao et al), 53

submarines, xiv, 4, 53, 75–76, 110, 161, 174. *See also* boomers

supercomputers, 102–3, 140. *See also* Singularity Computers

superintelligence race, 148–49

superpowers: characteristics of new, 111, 119, 136, 141; year 2050 ranking of, 142–46

superthermites, 59, 63

surveillance: drone, 61, 85, 87, 161; satellite, 70, 71, 148

swarming, 91–97

Swiss Federal Institute of Technology experiment, 149–50

synthetic chemistry, 25

Taiwan, 65, 195–98

tank armor, 30, 55, 59, 134

tanks, 59, 80–81, 134

T-cells, 86–87

technological life-forms, 184

technological singularities, 139–41. *See also* Singularity Computers

telecutters, 81

telephonelink between U.S. and Russia, 123

teleplanes, 81

teletanks, 80–81

terrorists: in future cold war, 120–21; scenarios on, 109–10, 123–33; targeting, 87–88

"There's Plenty of Room at the Bottom" (Feynman), 10

3-D printers, 140

Three Laws of Robotics, 151

3tags.org, 86

Tianhe-2 supercomputers, 102–3, 143

Tianhe-3 Singularity Computers, 143

top-down approach to building nanotechnology, 24, 25, 38–39

top secret classification of nanoweapons, xii, 8–9, 10, 45–46

torpedoes, 79–80

Torpex, 82–83

toxicity of nano-based products, 13–15, 35, 41

treaty extension to nanoweapons, 184–86, 187

Treaty of Brussels, 116, 144, 201

Treaty of Mutual Cooperation and Security, 117, 202

Treaty on the Non-Proliferation of Nuclear Weapons, 113, 116, 144, 164–65, 185, 187

Trinity atom bomb test, 15

Turing, Alan, 103

Turing test, 103, 104

UAVS (unmanned aerial vehicles). *See* drones

Umbrella Murder, 6–7

UN Commission for Conventional Armaments, 183

unintentional missile launches, 186

United Kingdom: future cold wars and, 119–20; nanoweapons and, 115–16, 143–44, 195, 198; nuclear weapons and, 113, 165; World War I weapons of, 80. *See also* Great Britain

United Nations: arms reduction and control and, 107, 146, 164, 179, 185–86; in scenarios, 174, 176–77; WMD definition of, 183

United States: nanotechnology arms race and, 5–6, 11, 12–13, 63, 111; nanotechnology spending of, 66–67, 194–95, 197; nanoweapon capabilities of, 45–46, 114, 142; overview of nuclear weapons of, 95; overview of past weaponry development of, 79–84. *See also specific agencies, events, and programs of*

"Universal Computing by DNA Origami Robots in a Living Animal," 96
University at Buffalo, 54
University of California Berkeley, 57
University of Massachusetts Amherst, 49–50
unmanned aerial vehicles (UAVS). *See* drones
unmanned helicopters, 84, 161
unmanned surface vehicles (USVS), 84
U.S. Air Force, 60–64, 82, 83, 161
U.S. Army, 55–60, 82–83, 84, 85, 160–62, 189–92
U.S. Army Posture Statement, 157
U.S. Army Research Office, 56
U.S. Energy Department, 111–12
U.S. Naval Research Laboratory, 50–51, 52
U.S. Navy, xiv; laser weapons and, 11–12; nanotechnology and, 47, 50–54, 160; torpedoes and, 79–80; unmanned vehicles use by, 83, 84
U.S. Patent and Trademark Office, 67
USS *Zumwalt*, 54
USVS (unmanned surface vehicles), 84
UV radiation, 147

V-1 cruise missiles, 81
V-2 ballistic missiles, 81
vacuums, 93–94
Vance, Marina E., 33
Vasilevich, Valery, 84
VB-1 Azons, 83
Vejerano, Eric P., 33
viruses, 19, 41, 92, 97, 134–35
vision for the blind, 140
Vita-More, Natasha, 29
von Clausewitz, Carl, 154

Wall Street Journal, 62
war: binary perception of, 153–54; defined, 153–54; deterrence of, 74, 146, 186; limited conventional, 154; truisms of, 43, 134; types of, 154–59. *See also specific wars and types of*
weapon delivery systems, 13, 74–75, 119, 136
weapons of choice, nanoweapons as, 118–19
weapons of mass destruction (WMD): autonomous smart nanobots as, 162, 187; defined, 183; nanoweapons classification as, 182, 183–85; literature on, 182; nuclear weapons as, 43. *See also specific types*
wearable nanomaterials, 35, 55, 56–57, 72, 160
weight reduction, 55, 56–57, 59
Whitesides, George M., 22
Wickersham Land Torpedoes, 80
WMD. *See* weapons of mass destruction (WMD)
Woodrow Wilson International Center for Scholars, 32
Work, Robert, 84
World War I, 80, 83, 134, 166
World War II, 13, 80–83, 111, 165
WU-14S, 63, 69–70, 75

year 2050 envisionment, 139–52; cold war and, 148–49; disruption in, 141–42, 146–47; new superpower ranking of, 141, 142–46; Singularity Computer military prowess and self preservation in, 149–51; summary of, 151–52; technological singularities of, 139–41
Yu-71S, 63

Other works by Louis A. Del Monte

The Artificial Intelligence Revolution: Will Artificial Intelligence Serve Us or Replace Us?

How to Time Travel: Explore the Science, Paradoxes, and Evidence

Unraveling the Universe's Mysteries: Explore Science's Most Baffling Mysteries, Including the Big Bang's Origin, Time Travel, Dark Energy, Humankind's Fate, and More